THE LEARNING AND TEACHING OF GEOMETRY IN SECONDARY SCHOOLS

IMPACT (Interweaving Mathematics Pedagogy and Content for Teaching) is an exciting new series of texts for teacher education which aims to advance the learning and teaching of mathematics by integrating mathematics content with the broader research and theoretical base of mathematics education.

The Learning and Teaching of Geometry in Secondary Schools reviews past and present research on the teaching and learning of geometry in secondary schools and proposes an approach for design research on secondary geometry instruction.

Areas covered include:

- teaching and learning secondary geometry through history;
- the representations of geometric figures;
- students' cognition in geometry;
- teacher knowledge, practice, and beliefs;
- teaching strategies, instructional improvement, and classroom interventions;
- research designs and problems for secondary geometry.

Drawing on a team of international authors, this new text will be essential reading for experienced teachers of mathematics, graduate students, curriculum developers, researchers, and all those interested in exploring students' study of geometry in secondary schools.

Patricio Herbst is Professor of Education and Mathematics at the University of Michigan, USA. Patricio is a former high school mathematics teacher in Argentina and his research focuses on the nature of the mathematical work that students do in secondary classrooms and the work that teachers do to manage knowledge development.

Taro Fujita is a lecturer in mathematics education at the University of Exeter, UK. Originally a mathematics teacher in Japan, Taro currently teaches the learning and teaching of mathematics in primary schools and higher mathematics, and he is also editorial assistant for the *International Journal for Technology in Mathematics Education*.

Stefan Halverscheid is Professor of Mathematics Education at the University of Göttingen, Germany. His background is in complex and differential geometry in the presence of symmetries and in research on the teaching and learning of mathematics. He has experience as a high school teacher, has lectured in teacher education at Münster, Oldenburg, and Bremen Universities, and was Dean of Studies and Dean of the Faculty of Mathematics and Computer Science at Göttingen.

Michael Weiss is currently on the faculty of the Department of Mathematics at the University of Michigan, USA. His background is in mathematics education and pure mathematics, and he was formerly a high school mathematics teacher in the United States.

IMPACT (Interweaving Mathematics Pedagogy and Content for Teaching)

The Learning and Teaching of Geometry in Secondary Schools
A modeling perspective
Patricio Herbst, Taro Fujita, Stefan Halverscheid, and Michael Weiss

The Learning and Teaching of Algebra
Ideas, insights and activities
Abraham Arcavi, Paul Drijvers, and Kaye Stacey

THE LEARNING AND TEACHING OF GEOMETRY IN SECONDARY SCHOOLS

A Modeling Perspective

Patricio Herbst,
Taro Fujita,
Stefan Halverscheid,
and Michael Weiss

Routledge
Taylor & Francis Group
LONDON AND NEW YORK

First published 2017
by Routledge
711 Third Avenue, New York, NY 10017

and by Routledge
2 Park Square, Milton Park, Abingdon, Oxon OX14 4RN

Routledge is an imprint of the Taylor & Francis Group, an informa business

British Library Cataloguing in Publication Data
A catalogue record for this book is available from the British Library

Library of Congress Cataloging in Publication Data
Names: Herbst, Pat.
Title: The learning and teaching of geometry in secondary schools : a modeling perspective / Pat Herbst, Taro Fujita, Stefan Halverscheid and Michael Weiss.
Description: New York NY : Routledge, 2017.
Identifiers: LCCN 2016044683 | ISBN 9780415856904 (hardback) | ISBN 9780415856911 (pbk.) | ISBN 9781315267593 (ebook)
Subjects: LCSH: Geometry—Study and teaching (Secondary)
Classification: LCC QA461 .H43 2017 | DDC 516.0071/2—dc23
LC record available at https://lccn.loc.gov/2016044683

ISBN: 978-0-415-85690-4 (hbk)
ISBN: 978-0-415-85691-1 (pbk)
ISBN: 978-1-315-26759-3 (ebk)

Typeset in Bembo and Stone Sans
by Swales & Willis Ltd, Exeter, Devon, UK

CONTENTS

IMPACT—SERIES FOREWORD

IMPACT (*Interweaving Mathematics Pedagogy And Content for Teaching)* is an exciting new series of textbooks dedicated to mathematics education and suitable for teacher education. The leading principle of the series is the integration of mathematics content with topics from research on mathematics learning and teaching. Elements from the history and the philosophy of mathematics, as well as curricular issues are integrated as appropriate.

Whereas in mathematics there are many textbook series representing internationally accepted canonical curricula, such a series has so far been lacking in mathematics education. It is the intention of IMPACT to fill this gap.

The books in the series will focus on fundamental conceptual understanding of the central ideas and relationships, while often compromising on the breadth of coverage. These central ideas and relationships will serve as organizers for the structure of each book. Beyond being an integrated presentation of the central ideas of mathematics and their learning and teaching, the volumes will serve as guides to further resources.

This volume treats the topic of geometry and is the second in the series—*The Learning and Teaching of Algebra* was the first. Geometry is a central topic in any high school mathematics curriculum around the world and a topic with inherent complexities due to factors such as the increasing numbers of students who are expected to study geometry and to opportunities for new ways of doing geometry provided by technological change. The volume presents a coherent view of the central ideas and relationships that integrates geometry content with the main issues and results from research, thereby making it a particularly timely choice as the second volume in the series.

Series Editors

Series Advisory Board

ACKNOWLEDGMENTS

This book resulted from a deliberate attempt to address an issue of international appeal, for an international audience of graduate students and their instructors. It took an international team to do so—among the four authors, we bring experiences of research into the teaching and learning of geometry in five countries (Argentina, Germany, Japan, the United Kingdom, and the United States). To create a semblance of a single authorial voice, we had to learn to write together: With the exception of the pre-existing collaboration between Michael and Pat, it was a novel experience for us to write with each other. The collaboration allowed us to rely on each other's expertise to produce something that nobody could have done individually, but it also required us to learn to work across truly different writing cultures. The language choices had to be negotiated between English, German, Japanese, and Spanish, and occasional sources in other languages such as French or Italian, in order to write something that was readable in English. But we also had to understand how our diverse academic backgrounds and writing cultures were reflected in the text we produced: While some of us tended to write compactly, others elaborated; while some alluded to what the reader might already know, others made no assumptions of the reader. We worked hard to maintain the collaboration: Rather than read each other like a critic or a reviewer, we used empathy and patience, we tried to meet each other's expectations and contribute what was missing in order to build on each other's strengths. The work has not been easy but inasmuch as it has been accomplished, it has made the four of us the first group of learners to take advantage of this book.

We say that not only to illustrate the benefits and challenges of cross-cultural collaboration but also to anticipate benefits and challenges that our international readers are bound to encounter in working through the book. It is very likely that the reader will have similar experiences with our text to those we had with each other's texts, and we recommend that the reader adopt a constructive stance: The real text will come to life in the interaction of the reader with the artifact he or she holds in his or her hands. The study of this book is not only an opportunity to learn about the teaching and learning of geometry, but also an opportunity to learn about different ways of reading and writing scholarship and, as a result, of reflecting on one's own reading and writing culture.

Therefore, we acknowledge each other for staying engaged with this project till its end. We also acknowledge the trust and guidance of our editors, Frank, Tommy, and Günther; as our editor designate, Frank was always available, resourceful, and wise. We acknowledge the comments of our reviewers Nathalie Sinclair, Keith Jones, Mike Battista, and Angel Gutiérrez. We were able to make use of most of their constructive comments; while we realize we may not have met all their expectations we remain indebted to the dedication they put into providing critiques and suggestions. We also used feedback on earlier drafts from our associates. In particular, Pat acknowledges the support of Nicolas Boileau in helping complete the reference list, and the comments from Nic as well as other members of the GRIP team at Michigan (Vu-Minh Chieu, Justin Dimmel, Ander Erickson, Umut Gürsel, Kristi Hanby, Mollee Huisinga, Inah Ko, and Amanda Milewski).

Inspiration and impetus to write this book draw from years of personal histories with geometry and its teaching and learning. Our former instructors, former co-teachers, research collaborators, and the many teachers and students from whom we have learned in our research have been present in our minds as we wrote the book. Finally, we acknowledge the support of our families. Pat thanks Vilma, as well as Ana Sofía and Bruno. Michael thanks Fruma, Isaac, Sarah, Kinneret, Chana, and Tova. Taro thanks Hannah, Emiko, Joh, and Tomo. And Stefan thanks Jana as well as Noah, Peer, and Lukas.

INTRODUCTION

"geometry is the art of reasoning well from badly-drawn figures"

Poincaré (1895)[1]

This book fleshes out a proposal for research and practice in the teaching and learning of geometry at the secondary level. We propose a possible way of thinking about events in students' study of geometry in secondary schools that addresses, in particular, what is being learned and what could be learned. This is a graduate textbook, and thus it demands some work on the part of the reader; but it is also an introductory textbook, in that it is proposed as a way to begin a research journey into scholarship in secondary geometry. It proposes a perspective on past and current research, and a perspective on possible improvements associated with the notion of modeling and the research that could be done within such perspective.

A question that has appeared every so often in the discourse of policy-making and curriculum development is: Why do we have students study geometry? In the United States, Herbst (2002a) has documented how, at the end of the nineteenth century, the curricular discourse included arguments as to whether and why to incorporate a range of new fields of study (e.g., the then-new sciences of biology and chemistry) to the secondary curriculum. Geometry had been a field of study from antiquity, and its place in the curriculum had been warranted on cultural transmission grounds. But when new fields of study began to claim a place in the curriculum, a different discourse of justification emerged. The Committee of Ten (National Education Association, 1894; Eliot, 1905) recommended the study of geometry on the basis of its capacity to develop students' logical reasoning capabilities, while other subjects were warranted on account of other roles disciplining mental faculties, such as observation capabilities. During the twentieth century, various other arguments emerged and coexisted along with that logical argument to justify the study of geometry. González and Herbst (2006) have identified four such arguments in the US: (a) a logical argument, derived from the efforts of the Committee of Ten, that justified geometry as an exercise in deductive logic; (b) an intuitive argument, that justified geometry as providing a language for relating to the experiential world; (c) a utilitarian argument, that justified geometry as providing tools for work in the real world; and (d) a mathematical argument, that justified geometry as providing experiences in doing the work that mathematicians do. Some of those arguments have recurred in other countries as well: In the United Kingdom, John Perry, Professor at the Royal College of Science, attacked the notion that geometry teaching should strictly be based on Euclid because "every Gentleman should know Greek thought" (Griffiths; 1998, p. 195), and instead advocated an engineer's point of view on geometry (Perry, 1902; Howson, 1982), stating that the usefulness of studying mathematics should be considered to "determine what subjects ought to be taught to children and in what ways" (Perry, 1902, p. 159). He severely criticized the emphasis on training students in the logical way of thinking in the teaching of mathematics (pp. 168–9). Instead of such orthodox methods, Perry insisted that it would be important for students to experience various mathematical activities and discover mathematical facts for themselves (p. 165). Similar reconsiderations of the teaching and learning of geometry at the beginning of the twentieth century also emerged in France, Germany, and even far east countries, as we document in Chapter 1.

For much of history, geometry itself had been conceived in relation to the Euclidean canon, even if the texts used to expound on it had included deviations from Euclid's *Elements*. Yet developments in the discipline starting in the seventeenth century and including first projective and non-Euclidean geometries, and later finite geometries, formal accounts of Euclidean geometry, and the algebraization of geometry, would over time bring with them the opportunity to tinker with the question of what geometry is. The secondary school curriculum has made some room for diverse conceptions of geometry: Although radical changes to the Euclidean canon have been rare and generally short-lived, the curriculum has incorporated bits and pieces of different ways of conceiving of the subject, likely in an effort to reconcile various compelling arguments for the study of geometry in secondary schools (see Cooper, 1985).

The twentieth century saw important educational reforms that brought with them curricular innovation; it also made room for the growth of research in mathematics education and related fields that attended in particular to geometry. In spite of calls from some mathematicians to eliminate Euclidean geometry from the school curriculum on the grounds of its supposedly outdated role in the discipline (Fehr, 1963), Euclidean geometry has continued to be a mainstay of secondary education. And yet many things have changed in the world outside of school, suggesting the need to revisit the question: Why study geometry? As we progress into the twenty-first century, developments in technology and in the role of mathematics in life and work have brought with them the need to reevaluate the question of what geometry literacy the educated citizen needs. Technological progress has brought us computing technologies, digital graphics, and user interfaces that afford unprecedented ways of interacting with representations of geometric ideas on the screen and elsewhere (consider 3D sketching software and 3D printers); technology has also brought possibilities of visualization and manipulation that enable us to experience and record our interactions with the material world (consider motion capture, drones, and digital video). In general, present day technology permits us to consider geometry instruction as an exercise in multimodal[2] modeling of the physical world. The expansion of the practice of mathematical modeling to all aspects of human life has made almost any object or practice in the everyday world amenable to progressive understanding and prediction in ways that can be mathematically mediated. These possibilities bear important relation to the question of why young people should study geometry in secondary school.

It is therefore fitting to start this account by taking a position on the question of what geometry is and why it should be taught in secondary schools. We do that here, not expecting that this position will account for current research or practice, but rather expecting that it will propose a way of thinking about research and practice that can be generative and useful for graduate students in mathematics education as they start working in research and development.

We start from the assumption that elementary school and the physical, computer-based, and social activities of children during the elementary school years have afforded them diverse experiences with space and shape that they bring with them when they arrive in secondary schools. Those earlier experiences have involved them in proximal contact with a variety of artifacts, including words, diagrams, scale models, photographs, and videos of objects that, from the perspective of the mathematically educated, are amenable to being described as geometric representations—that is, as things that stand for or refer to geometric concepts. Based on that assumption, we propose that the development of geometric knowledge at the secondary level consists of the progressive sophistication of students' intellectual means to model, predict, and control geometric representations—that is, the progressive sophistication in students' ways of organizing those artifacts (words, diagrams, and others) so that they can be reliably used in making and transacting meanings. Said in a different way, while elementary school geometry introduces children to the symbol systems for representing the human experience with shapes and space, including the production of words, diagrams, and other artifacts, secondary school geometry can help students develop models and theories and use them to predict and control the variance among those. Such progressively sophisticated models and theories constitute the core of secondary geometry. They relate to the geometries of mathematicians, for example as explicated by Hilbert in his axiomatic treatment of Euclidean geometry (Hilbert, 1902), as elements of a convergent sequence relate to its limit.

Hence we see the study of geometry in secondary school as the development of the capacity to intellectually organize, predict, and control the world of representations of physical objects and experiences. We describe this as a *modeling perspective* insofar as it sees as central the process of creating models[3] of diagrams and other concrete representations, models that aspire to being used to predict and control

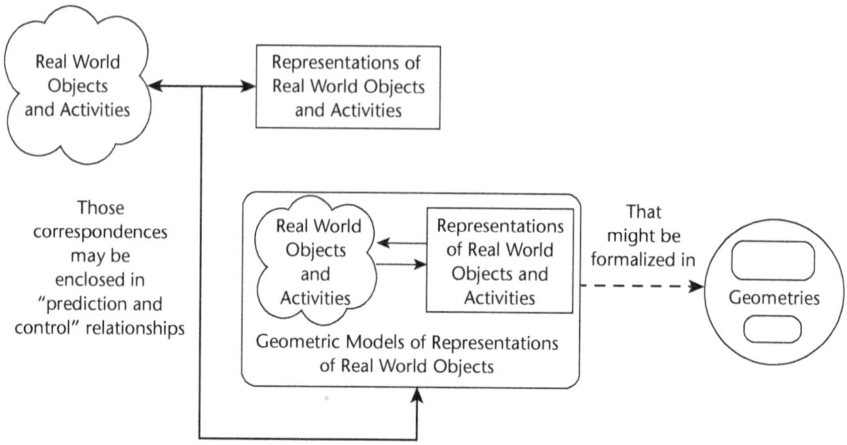

FIGURE 0.1 Secondary geometry as a source of models

those diagrams and other concrete representations (see Jahnke, 2007). Figure 0.1 may help explain this perspective. Reading from left to right, the first arrow in the diagram depicts how, in elementary school mathematics and other out-of-school activities such as photography or art, children create representations of real world objects and activities as means of recording or reporting on them. The second arrow takes that representational activity and puts it inside a modeling environment, which we propose is the key activity of secondary geometry: We expect that secondary school geometry will engage students in the prediction and control of those representations, using the real-world experiences as grounds and targets of such prediction, thus treating those representations as geometric models. The

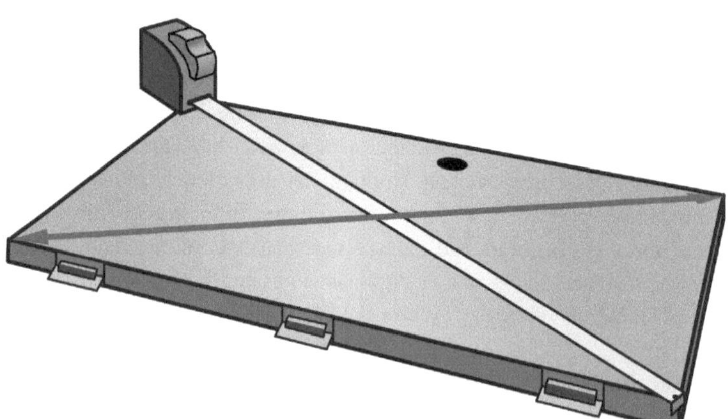

FIGURE 0.2 How to check that a door's angles are right

dotted arrow suggests that further work formalizing those models, for example by proposing an axiomatic structure, might converge to a formal geometry such as Hilbert's. Because the words *representation* and *model* are often used interchangeably, it is worth clarifying here that we use *representation* to describe the role of a physical or otherwise concrete artifact in communication as something that points to something else, while we use *model* as a set of relationships among representations, itself represented in progressively abstract means (e.g., at some point with mathematical symbols and logical relationships). While the student relates to the first by way of perception and actual transaction (e.g., talk), the student relates to the second by conception and intellectual means of control, including calculation. But the distinction is not sharp; rather, the distinction tries to express, albeit in a simplified way, two extremes of a continuous process of mathematization.

A simple example may help ground this distinction and illustrate this perspective. The diagram shown in Figure 0.2 shows how to check that a door is rectangular by providing instructions to measure its two diagonals. It is expected that elementary school students would have no trouble recognizing this as the picture of a door, and that they would have learned to do measurements such as those indicated in the figure; they might thus be able to relate to those two measures as numbers that might turn out to be equal. Elementary school students might also have learned to associate the word rectangle to the door. This is an incipient example of modeling, where *rectangle* as a mathematical term describes the door. Elementary school students are likely to have more of that descriptive modeling knowledge: For example, they might relate the measures obtained to the diagonals of a rectangle. An important change in perspective would appear in secondary school, enriching that capacity to understand the diagram: For example, it would equip students with the awareness that, inasmuch as the door is a rectangle, the measured diagonals *should be* congruent; reciprocally, if they turned out not to be congruent, students may be disposed to claim that the door might not be a rectangle after all. Furthermore, as the door is represented in perspective, secondary geometry would equip students to consider that minor differences in the actual length of the representation of the diagonals in the photograph should not suggest that the actual door is not rectangular (but note that Figure 0.3 presents a very different case). We would expect that secondary geometry would equip students with the knowledge needed to justify why the measures of diagonals of a rectangular door should be equal (e.g., by way of triangle congruence or the Pythagorean theorem). Further, we might expect that secondary geometry would equip students with the knowledge needed to expect that diagonals of a rectangle seen in perspective might not appear congruent, yet also be able to tell a rectangle apart from a non-rectangle based on a picture even if neither of them looks like a rectangle (see Figure 0.3). Such analysis might combine symbolic and iconic elements—either speaking of sides and diagonals or drawing lines on the photograph. The modeling language need not always be symbolic, but it would in general permit the increase of calculation and logical inference (or warrantable prediction, in general) in the production of knowledge.

FIGURE 0.3 Which of the two quadrilaterals is more likely to be a rectangle?

The book reviews historical and epistemological aspects of the curriculum (in Chapters 1 and 2), issues of thinking and learning (in Chapter 3), and of teaching and teachers' knowledge (in Chapter 4), in order to land on a set of proposals for geometry instruction (in Chapter 5) that can be addressed in a design research agenda on this modeling approach (in Chapter 6). Those proposals are provided as seeds for projects that graduate students might take on in the context of action research or design-based research. They all derive inspiration from the notion that geometry education can be the development of the intellectual means to model diagrams and other concrete representations; further they nest that notion in the context of learning situations that involve students in communication about information supported by such diagrams and concrete representations. The notions of modeling and communication are thus the pillars on which the book is supported.

Each of the chapters serves two functions. On the one hand the chapters review for the reader important points that the field of mathematics education has learned from research in geometry instruction. On the other hand, they position the reader to undertake independent research projects of an instructional nature. We imagine our readers as Masters or doctoral students interested in deepening their knowledge of the subject with the goal of being able to pose research questions and undertake research projects that connect to the teaching and learning of geometry. We believe this book might provide them with a big-picture vision of why geometry should continue be taught in the twenty-first century. While we don't completely flesh out that vision, the various instructional interventions scoped in Chapter 5 are not only contexts for research projects, but also examples of what that vision might look like.

Notes

1 Cited in Bartocci's (2013) article on the birth of algebraic topology. Originally from Poincaré, H.: Analysis situs. Journal de l'École Polytechnique, 2ème série, 1, 1–121 (1895) (reprinted in Œuvres, vol. VI, pp. 193–288.) English trans. Papers on Topology: Analysis Situs and its Five Supplements, John Stillwell, trans. Washington D.C. and London, American Mathematical Society and London Mathematical Society (2010).
2 The multimodal qualifier suggests that modeling might be done with a variety of semiotic resources, for example diagrams as well as symbols.
3 By a *model* we mean an operational representation, or a representation made in a representation system that includes (implicit or explicit) rules that enable calculation with representations, thus producing other artifacts that might be interpretable as representations themselves. To connect it with a model that may be better known to the reader, consider the following non-geometric example: A regression equation models the relationship between two random variables not only because it represents data known about the variables but also because it is contained within a system of rules for calculating (predicting) new values. By *modeling* we mean the disposition to and the practice of creating progressively more explicit and precise operational representations.

1

THE DISCOURSE OF TEACHING AND LEARNING SECONDARY GEOMETRY THROUGH HISTORY

1.1. Introduction

On a winter morning in 1959, one of the twentieth century's greatest French mathematicians stood before the leading mathematicians, educational policy makers, and secondary mathematics teachers of 18 nations, and declared war on a text that had stood at the center of the traditional mathematics curriculum for over 2,000 years. At the Royaumont Seminar, convened in Asnières-sur-Oise by the Organisation for European Economic Co-operation (OEEC), Jean Dieudonné—a key member of the pseudonymous Bourbaki collective, who also wrote in the late 1950s on the teaching of geometry at school—announced: "*A bas Euclide! Mort aux triangles!*" ("Down with Euclid! Death to triangles!") (Roberts, 2006, p. 157; see also Fried, 2014). In the subsequently published version of Dieudonné's address (Dieudonné, 1959) this call was presented in the slightly more moderate form "Euclid must go!" (p. 35), but the critique was no less pointed. Despite claiming to hold the "deepest admiration for the achievement of the Greeks in mathematics", Dieudonné nevertheless described Euclid's geometry as little more than "a chaotic heap of results with no significance except as scattered relics [. . .] or an obsolete approach" (p. 35). The school curriculum, argued Dieudonné, needed to be purged of the "dead-weight of 'pure geometry'" (p. 38) to make space for a "modern" curriculum.

Dieudonné's call for revolution influenced much of the discourse in mathematics education throughout the "New Math" era of the 1960s, and the ways in which students studied and learned geometry did indeed undergo profound transformations over the subsequent decades. Hansen (1998) refers to the ensuing changes in the curriculum (particularly in Europe) as a "downgrading of the teaching of geometry at all school levels" resulting from "the algebraization and formalization of mathematics in school" (p. 235). Notwithstanding this significant change in the role of geometry in school though, Euclid—perhaps needless to say—has not gone away. Despite decades of curricular innovation and multiple waves of reform efforts both within and across national boundaries, the geometry of Euclid continues to play a significant role in the secondary mathematics curriculum of virtually every country.

What was the background to Dieudonné's call for revolution? And why does his proposal still resonate today, more than half a century later? This chapter tries to identify the forces that have shaped the geometry curriculum, instruction, and learning from a historical and an international point of view. The discourse of teaching and learning geometry over the centuries provides a rich, at times unexpected, source of examples and indicators of why and how geometry can be taught and learned, and document which decisions regarding the teaching and learning of geometry can and must be made.

1.2. Overview of This Chapter

To structure the discussion in this chapter, we make use of how historians and mathematics educators have identified different reasons for curricular developments. Hansen (1998) starts from the observation that curricular debates, in general, must take into account four different perspectives:

- the *epistemological* perspective, which sketches the impact of the development of geometry into a scientific, logically coherent discipline (Glymour, 1977) on its teaching and learning;
- the *pedagogical* perspective, which focuses on central developments in the history of teaching geometry (Barbin and Menghini, 2014), from the design of textbooks to assuming the perspective of individual learners;
- the *technological* perspective, which includes such subjects as the influence of dynamic geometry on mathematics curricula; and
- the *political* perspective, which considers the political, cultural, and societal settings in which school curricula are developed (Damerow and Westbury, 1985).

Of course, the teaching and learning of geometry goes back much further than compulsory school attendance and curricula in mathematics. The study of geometry in its practical applications to agriculture, construction, and astronomy is indeed virtually as old as civilization itself. In this book, we are interested primarily in the teaching and learning of geometry in settings that resemble schools as we know them, and for that reason our narrative will focus largely on developments since the nineteenth century. Looking more specifically at the case of school geometry, it seems appropriate to follow variations of these perspectives according to three historical periods, demarcated by the late nineteenth century's so-called "foundational crisis" on the parallel axiom and by the twentieth century's introduction of the computer in the classroom. As we will see below, the foundational crisis was more of an extended period than a single event: Even before geometry was reestablished on a rigorous foundation in the early twentieth century, new, influential geometry courses had already begun to appear and shape the way geometry was taught and understood. The presence of computers in the classroom makes the third historical period reflect more recent discussions on curricular developments in view of the options that dynamic geometry software provides.

In their account of the history of teaching geometry, Barbin and Menghini (2014) identify two different aspects of geometry: The "deductive/rational" and the "pseudo-practical/intuitive" approach, which built up a "tension" that "manifested itself many times throughout history" (p. 473). These aspects structure our considerations of the teaching and learning of geometry till the resolution of the crisis on the parallel axiom. Curricula in mathematics, in the modern sense, began to appear in the second half of the nineteenth century. Hansen (1998) identified five objectives as most important to the shaping of geometry curricula: To "establish knowledge", to "prepare [. . .] for applications", to present classical "milestones in the development of geometry", to "develop skills and abilities", and to "strengthen logical thinking and deductive reasoning" (p. 236).

Apart from the interest in milestones, these objectives correspond approximately with the four modal arguments identified by González and Herbst (2006), who studied the justifications put forward in the United States between 1890 and 1980 for why geometry should be studied in secondary schools: (1) a *mathematical* argument, which stressed that geometry could provide a context for students to experience the kinds of activities that are typical of mathematicians; (2) a *formal* argument, based on the claim that geometry trains students in deductive logic; (3) an *intuitive* argument, which emphasized the power of geometry to provide vocabulary for naming, and experiences for exploring, the real world; and (4) a *utilitarian* argument, which stressed the geometry that is especially useful in students' future working worlds.

With respect to more recent curricular developments of geometry, Galuzzi and Neubrand (1998) point to the following influential areas: (1) changes in the development of mathematics itself, (2) greater emphasis on applications and modelling, (3) the debate over fundamental ideas, (4) constructivist ideas of learning, (5) the focus on mathematics as a human activity, (6) geometry seen by students as an empirical theory, and finally (7) the impact of the computer. In the fourth section of this chapter, these areas are investigated from an international perspective by re-examining nine countries whose geometry curricula have been previously compared (Hoyles et al., 2002). Since all these countries—Canada, Great Britain, France, Germany, the Netherlands, Japan, Poland, Singapore, and Switzerland—have revised their curricula since Hoyles et al's (2002) comparison, some of those more recent developments are described and the main differences among them are articulated. In the context of that comparison, we give special attention to the options dynamic geometry software affords for the teaching and learning of geometry by discussing its impact on the development of the curricula of the countries mentioned above.

1.3. The Development of Geometry up to the So-Called Foundational Crisis

What *is* geometry? The subject of geometry is vast, and related to many phenomena both inside and outside mathematics. This section takes stock of major themes and developments in the subject matter as they concern in particular its teaching and learning. The chapter does not aim to cover the long history of geometry; we refer to Barbin and Menghini (2014) for a systematic review of the history of the teaching and learning of geometry. But for their central role in this book, we deal in particular with the history of reasoning and figural concepts.

1.3.1. The History of Reasoning and Figural Concepts

Applications in craftsmanship and astronomy were the source of pragmatic, hands-on geometric rules since the beginning of highly developed civilizations in ancient Egypt, Arabia, and India. Several facts and rules of geometry are attributed to the ancient Greeks, probably because of their elaborate educational system. According to Eudemus, Thales (around 580 BCE) is said to have discovered and formulated important geometric results: "(1) that a circle is bisected by its diameter; (2) that the angles at the base of an isosceles triangle are equal [. . .], (3) that when two straight lines intersect, the vertically opposite angles are equal; (4) that if two triangles have two angles and one side equal, the triangles are equal in all respects." (Dicks, 1959, p. 302). Over the course of the next few centuries, philosophical schools linked research, tradition, and education. For geometry, Hippasus of Metapontum (a follower of Pythagoras' school between 530 and 450 BCE) codified results that had already been partly known to the Babylonians (approximately 1600–1300 BCE) as utilitarian facts; of those results, the Pythagorean theorem is surely the most important example (see Eves, 1990). The idea of a demonstration or proof gradually became an integral part of philosophical reasoning and ultimately led to the development of important mathematical concepts, such as commensurability and irrationality. However, Szabo's thesis that the idea of a proof is an invention of "Greek mathematics" (Szabo, 1978) has been contested, for instance by Waschkies (1989).

Geometry was considered particularly important in the philosophical schools of ancient Greece. Plato's theory of ideal forms shows the influence of geometry on philosophical thought more broadly. Geometry was also closely connected to teaching and learning: The dictum "Let no one ignorant of geometry enter here" is said to have been inscribed over the door of Plato's Academy (Smith, 1999, p. 131). In Plato's *Meno*, Socrates engages Meno's slave in a discussion of the problem of doubling the area of a square as an illustration of Socrates' view of the origins of knowledge. The episode is a famous example of what has come to be known as the Socratic method, and also shows some features that are typical in teaching and learning of geometry. In particular, Plato's dialogue depicts the activity of *hypothesizing* in the course of defining and solving problems (Wyller, 1964).

According to Plato, geometric "shapes had an independent life in the world of ideas" (Hansen, 1998, p. 12). By focusing further attention on the distinction between the world of ideas and the world of experience and sensation, Aristotle played a crucial role in the development of a foundation for mathematics. The philosophical concept of the world of ideas made it possible to develop an abstract concept of shape: "The concrete figures live in the real world and the abstract in the world of ideas" (Hansen, 1998, p. 12). That is to say, the triangles and circles whose properties are studied in geometry were understood to be distinct from the physical and diagrammatic representations of them; the latter were regarded as imperfect realizations of the pure abstract forms that were the proper focus of mathematical study. This marked the beginning of mathematical thinking about objects that exist solely in the world of ideas, and the ontological status of such abstract objects was assured solely by the soundness of the reasoning that was applied to them:

> From the perspective of ontology or of epistemology the question may be asked: 'What is the nature of mathematical objects?' To ask the question ontologically amounts to asking for the real subjects, the things in the world, with which mathematics deals. Epistemologically the question is more likely to be directed at mathematical reasoning: 'What is mathematical reasoning about?'
>
> Plato seems to have given the same answer to the question from both perspectives. It is even possible that his answer to the ontological question was inferred from his epistemological analysis of mathematics in some way like the following: Mathematicians reason as if they were dealing with objects that are different from all sensible things, perfectly fulfill given conditions, and are apprehensible by pure thought; mathematics is correct; therefore, there are such objects. Argument of this kind is also characteristic of the modern mathematical Platonist.
>
> *(Mueller, 1970, p. 156)*

Aristotle launched a systematic programme of organizing mathematics as a system based on definitions and postulates, and theorems deductively derived from that basis. About a century later, Euclid's *Elements* appeared as a compilation of previous works in geometry and number theory that had originated in the philosophical schools. The *Elements* is the oldest conserved systematic text on mathematics, and its impact in both shaping and following the ideal of thoroughly axiomatic reasoning cannot be overstated (Mueller, 1991). The Euclidean practice of deducing everything from a relatively small set of axioms, postulates, and definitions is often regarded as the preeminent example of Aristotle's methodology. However, there are significant differences between the use of terms in Aristotle and Euclid (Szabo, 1974) and in the logical organization of the *Elements* (Torretti, 1978, p. 5), which might indicate that the philosophical school or schools from which Euclid's *Elements* originated did not follow Aristotle's concepts from the beginning. In addition, the *Elements* can be (and indeed has been) critiqued as an imperfect execution of Aristotle's vision, insofar as it contains numerous tacit assumptions and other logical gaps. Some of these gaps may stem from the fact that the *Elements* drew on independent earlier sources whose contents and organization did not always fit neatly together; others may have their source in the inherent difficulty of studying the properties of geometric figures without relying on visually obvious features of those diagrams. Indeed, it was not until the early 20th century that a fully rigorous theory of geometry redressed the perceived flaws in Euclid's *Elements*.

In his commentary to the first book of Euclid, written about 80 years after its appearance, Proclus (cited as Proclus, 1970) mentions treatises of other philosophical schools. They were lost over the centuries and surely the continued existence of the *Elements* is an important reason for its influence. For a recent overview on research on who actually wrote which part of the *Elements*, what the original state of the work was, and whether the first aim of this work was actually pedagogical, see Bernard, Proust and Ross (2014). As a compilation of knowledge, the *Elements* treat a broad range of geometry and number theory: Books I–VI are on plane geometry, Books VII–X on number theory, and Books XI–XIII on solid geometry. The influence of the *Elements* on the curriculum is still felt today, insofar as many take the contents of the first six books as bounding the scope of what belongs in the geometry curriculum (Hartshorne, 2000b).

Equally if not more important is the influence of the *Elements* in shaping typical ways of working geometrically. Stamper (1909) identifies the following three principles of Euclid's *Elements*: "(1) Hypothetical constructions are not admitted. (2) All practical work is excluded. (3) All constructions are by means of straightedge and compass only. This bars out the conic sections." (p. 31). By *hypothetical constructions*, Stamper means the use of figures or auxiliary objects that have not been previously shown to be constructible with compass and straightedge. For example, Euclid does not make use of an angle bisector as an auxiliary object in a proof until after the construction of an angle bisector by compass and straightedge has been demonstrated to be possible (Book I, Prop. 9). It has been suggested that this strict avoidance of hypothetical constructions made it impossible for Euclid and his successors to discover results such as Morley's Trisector Theorem: Since there is no general procedure for constructing angle trisectors with compass and straightedge, without hypothetical constructions one cannot even begin to entertain questions such as what figure is formed by the intersection of the angle trisectors in a triangle (Swicegood, 2004, p. 42).

Constructions thus play a role in Euclid analogous to, but at the same time more restrictive than, that of existence proofs in modern mathematics. Indeed, the central role of constructions in the organization of the *Elements* and subsequent treatments of geometry is one of the key characteristics of so-called *synthetic geometry*. In many modern treatments of geometry, the idea of *geometric transformations* has, to a large extent, displaced the central role of constructions, although constructions remain very important: In two-dimensional settings, every figure of Euclid is constructible with straightedge and compass (Hartshorne, 2000a, p. 18).

Euclid's *Elements* suffers not only because of the presence of what, from our modern perspective, seem to be logical gaps, but also from complications arising from the state of mathematics at the time. Just to give an example: Since numbers in classical Greek mathematics were understood to be only what we would today refer to as positive integers, Euclid's *Elements* does not make reference to the measurement of line segments except in the special case when one segment exactly *measures* (that is, fits a whole number of times into) another segment; for this reason, instead of speaking about segments of *equal length*, Euclid develops the concept of *congruent* segments. Congruence still lacked a precise definition—and the mathematics of the time did not yet include the modern notion that undefined terms are fundamental building blocks of a mathematical language—but nevertheless the use of congruence did allow for the comparison of the magnitude of segments, the development of an algorithm for finding a greatest common measure for two segments (assuming that they were commensurable), and consequently the development of the notion of incommensurable magnitudes. This in turn made it possible to discuss what we would describe in modern terminology as the irrationality of $\sqrt{2}$, but which Euclid described as the incommensurability of a side of a square and its diagonal (Hartshorne, 2000a).

Despite its lack of modern-day rigor and vocabulary, the most characteristic feature of the *Elements* is its proofs. Three particular theorems play a central role as tools to develop the proofs in Euclid's elementary geometry (Artmann, 1999):

- *Elements*, Book VI, Proposition 2 ("If a straight line is drawn parallel to one of the sides of a triangle, then it cuts the sides of the triangle proportionally; and, if the sides of the triangle are cut proportionally, the line joining the points of section is parallel to the remaining side of the triangle"; Euclid, 1956, v.2, p. 194) connects parallelism with the ratios of segments.
- The Pythagorean Theorem and its converse, as presented in *Elements*, Book I, Proposition 47 ("In right-angled triangles the square on the side subtending the right angle is equal to the squares on the sides containing the right angle"; Euclid, 1956, v.1, p. 349) and Book I, Proposition 48 ("If in a triangle the square on one of the sides be equal to the squares on the remaining two sides of the triangle, the angle contained by the remaining two sides of the triangle is right"; Euclid, 1956, v.1, p. 368), connect right triangles with areas.
- The inscribed angle theorem in Elements, Book III, Proposition 20 ("In a circle the angle at the centre is double of [sic] the angle at the circumference when the angles have the same circumference as base"; Euclid, 1956, v.2, p. 46) provides a link between angles and concyclicity.

A century after Euclid, Apollonius of Perga wrote the *Conics*, and Archimedes of Syracuse developed the method of exhaustion as a means to prove geometric theorems about the area of a circle, the surface area and volume of a sphere, and the area under a parabola. These important contributions have found their way into geometry alongside Euclid's *Elements*. Still, it is not too much of an overstatement to argue that the monumental character of Euclid's accomplishment was so influential that it brought the development of geometry, in Europe at least, to a virtual standstill for about 2,000 years. (Although a full discussion is well beyond the scope of this chapter it is important to note that Hindu, Arabic, and Persian mathematicians remained active over this long period. It is also worth noting, with Netz, 1998a, p. 37, that the extent of Euclid's longstanding influence may owe more to the attention of the scholars of the Middle Ages than to his real importance in antiquity, which pales in comparison with that of Archimedes; see also Netz, 2002). Geometry might have developed much differently, with less of a separation between proof and application, for example, if Archimedes' work had received the attention that Euclid's Elements did (see Netz, 1998b; Nathalie Sinclair, personal communication, April 2016).

1.3.2. The Mathematical Emancipation of Euclid: From the Enlightenment to the Foundational Crisis

In the Enlightenment, mathematical developments were seen first in other areas, particularly in arithmetic and algebra. Descartes' (1637) *La Géométrie* brought important innovation to geometric reasoning by arithmetizing geometric objects: Descartes used letters x and y to represent the lengths of unknown line segments, and letters a, b, c, to represent the lengths of known line segments, and showed how algebraic manipulation of expressions involving those quantities enabled the solution of geometric problems. This fusion of algebra with geometry transformed mathematicians' conceptions of what geometry was and how it could function, and laid the foundation for a later rigorous formalization of axioms and proofs. Descartes' use of algebraic methods to solve geometric problems was taken up in school books such as Lamy's *Eléments de géométrie ou de la mesure des corps*, in different editions from 1685 to 1758, and laid the groundwork for the development of the infinitesimal calculus in the mid-seventeenth century (Barbin & Menghini, 2014).

Still, despite these new perspectives and methods in geometry, the *Elements* has had, and continues to exert, an enormous influence on the teaching and learning of mathematics in general and geometry in particular. Throughout Europe, there were numerous attempts, beginning in the late sixteenth century, to craft new textbooks, more suitable for school use but still organized essentially around Euclid's content. As examples, in Germany there were the texts of Clavius, Sturm, and Kaestner; in Switzerland, that of Scheubel; in France those of Pardies, Dechalles, and Tacquets; in England, Playfair and Todhunter; and in Russia, Fuss. In the nineteenth and early twentieth centuries, even though the *Elements* was typically studied at university level, some high schools continued to use his treatise as the main textbook for studying geometry. The *Elements* and its many successors shaped the very idea of textbooks for learning geometry within both universities and secondary schools, even long before compulsory school attendance and school systems (in the modern sense) were established in most nation-states.

One of the more striking features of Euclid's systematization of geometry is his inclusion of the 'parallel postulate.' In Euclid's formulation, this is postulate 5, which states that "if a straight line falling across two other straight lines makes internal angles on the same side less than two right angles, then, if produced infinitely, the two straight lines meet on that side" (Euclid, 1956, p. 202). The complexity of this postulate, especially in comparison with the other, more naturally self-evident postulates, combined with Euclid's apparent reluctance to make use of the parallel postulate (it is not used in any proofs until Book I, Prop. 29), has motivated inquiry into the nature of this property, and in particular into the question of whether it can be logically deduced from Euclid's other postulates. In the nineteenth century, this inquiry led to the discovery of non-Euclidean geometries, which ultimately gave rise to a crisis in the foundations of mathematics, linked to long-held uncertainty about the relationship of geometry to the real world of physical experience and to whether it was possible to definitively establish the consistency (i.e., the absence of contradictions) of any system of axioms for geometry.

There had been numerous attempts to prove Euclid's parallel postulate. Saccheri attempted (around 1700) to prove the parallel postulate by assuming, contrary to what may be proved from the parallel postulate, that the sum of the angles in a triangle is less than 180°. Saccheri claimed that the resulting theory was "repugnant to the nature of straight lines" (Saccheri, 2014, p. 45), and that therefore the parallel postulate must be true, but Gauss, Bolyai, and Lobatschevsky independently came to the conclusions that Saccheri's hypothesis does not lead to an actual contradiction. Several mathematicians (notably Legendre) persisted in attempts to prove Euclid's parallel postulate.

It was not until the late 1860s and early 1870s that mathematicians in general began to accept that the parallel postulate was necessary to characterize plane geometry, in the sense that there could be logically constructed and mathematically consistent non-Euclidean geometric models in which the postulate was false (Bonola, 1955). Discovered in 1868 by Beltrami, the pseudo-sphere was a differential geometric (local) example. In 1871, Klein worked out the links between non-Euclidean geometry and the projective metrics that Cayley had introduced in 1859. This established the existence of a model that satisfies all of Euclid's axioms except the parallel postulate: The hyperbolic plane. Through the function-theoretic model of Poincaré in 1880, hyperbolic geometry was finally established and showed the existence of a geometry that satisfies all of Euclid's axioms but the axiom of parallel lines.

However, the problem with Euclid (from the perspective of logical rigor) was not that the *Elements* had too many postulates, but that it had too few. Euclid had taken for granted numerous properties of the plane that are visually and intuitively obvious but never explicitly articulated; for example, the fact that given three distinct points on a line it is always true that one point lies *between* the other two. Scholars in the nineteenth century, notably Pasch, worked to bring explicit attention to those tacit propositions, including those of separation of the plane. The project of re-establishing Euclid's *Elements* on a rigorous logical foundation was finally completed with the publication of David Hilbert's *Grundlagen der Geometrie* (1899), published in English as *Foundations of Geometry* in 1902. Fully incorporating the recent discovery of non-Euclidean geometries, Hilbert's *Grundlagen* was noteworthy for its complete detachment from any ontological commitment. That is, Hilbert took as a basic principle that a mathematical theory is developed independently of the substantive interpretation of the basic concepts: Only in expressly formulated axioms are logical relationships important. Thus for Hilbert, the words *point, line, plane, congruent, between*, and so forth were taken as undefined terms. Hilbert is famously (though possibly apocryphally) quoted as having said "One must be able to say 'tables, chairs, and beer mugs' each time in place of 'points, straight lines, and planes" (Blumenthal, cited by Grattan-Guinness, 1998, p. 208). Less famously, but with clearer attribution, Hilbert wrote: "If I subsume under my points arbitrary systems of things, e.g. the system: love, law, chimney sweep. . . and then just assume all my axioms as relationships among these things, then my theorems, e.g. also the Pythagorean theorem, are true of these things, too." (Cited in Frege, 1980, p. 13).

Hilbert's approach to mathematical formalism paved the way for a modern treatment of mathematical structures. After important preliminary work, in particular by Pasch (1882), who closed gaps in Euclid's axiomatic system, Hilbert succeeded in his *Grundlagen* in characterizing coordinates with the help of geometric axioms. This means that a suitable collection of geometric postulates—including variations on those of Euclid, but extending that system to include axioms of betweenness, continuity, and so on—could completely determine a realization in a coordinate system like that of Descartes (Hartshorne, 2000b). In particular, Hilbert worked out the first fully axiomatic definition of real numbers. This gives another justification for his approach: If the theory of the real numbers is assumed to be consistent, then so too must be Hilbert's geometric theory. Interestingly, Hilbert originally wrote his *Foundations of Geometry* for a seminar for teachers.

Having set forth a completely rigorous foundation for geometry, Hilbert next set his sights on a similarly rigorous axiomatic organization of set theory (and indeed all of mathematics) together with a proof of its consistency. This ambitious goal, articulated in the early 1920s, came to be known as Hilbert's Program. Although extremely influential, Hilbert's Program came to an end when Kurt Gödel proved in 1931 that any theory in first-order logic that is strong enough to formulate the properties of arithmetic will inevitably be unable to resolve the truth status of certain statements that can be formulated in that theory, and in particular that no such theory can prove its own consistency; the relative consistency of a theory (i.e., that if theory A is consistent, then theory B is consistent as well) is all one can hope to prove. Gödel's incompleteness theorems put an end to research activities seeking new and better axiomatic foundations for geometry, but for mathematicians and educators, the problem remained how to write textbooks for teaching and learning geometry that are both appropriate for students and scientifically based.

The discovery of non-Euclidean geometry and the development of algebra and analysis led, inevitably, to a reconsideration of the role of Euclid's *Elements* (and its imitators and successors) in schools. If Euclidean geometry was only one of several mathematically consistent geometric systems, did the *Elements* deserve its privileged role in the curriculum? Indirectly but inexorably, these considerations led to the calls for reform of the curriculum in the mid-twentieth century, culminating in Dieudonné's denunciation of the *Elements:* "Euclid must go!"

1.4. The Shaping of Geometry Curricula in the Nineteenth and Twentieth Centuries

Euclidean geometry, in the form of a modified version of the body of knowledge first exposed in Euclid's *Elements*, has been taught for centuries, first in universities and later in secondary schools. For example, in the early 1800s in the US, geometry became part of the curriculum of the academies that prepared students for college (Herbst, 2002a). The reasons why geometry was taught are quite instructive. We recall the four modal arguments which González and Herbst (2006) identified in debates on the purpose of the study of geometry in the United States in the late nineteenth through twentieth centuries: Namely, that geometry:

(1) affords students experiences that resemble the work of mathematicians,
(2) provides students an opportunity to learn logic,
(3) helps students to interpret and systematize their intuition about the world,
(4) allows practical connections to the real-world skills students will need in their future working lives.

In what follows, we use these four modal arguments as frames for considering some of the ways in which the content of geometry —both as a mathematical body of knowledge and as a curriculum—was radically reconstructed in the nineteenth and twentieth centuries. In doing so we realize we are taking these four modal arguments rather far from the use for which they were intended; here we use them not to describe arguments for why students should study and learn geometry, but rather as a framework to consider the reconstruction and reorganization of the geometry curriculum (and indeed much of mathematics) in the decades following the discovery of non-Euclidean geometry.

1.4.1. Geometry as an Opportunity for Students to Experience the Work of Mathematicians

To approach the first of the four modal arguments identified by González and Herbst (2006) we may well ask, then: What are some of the characteristic activities of mathematicians, and how has the geometry curriculum accommodated these activities?

One essential aspect of mathematical work is the seeking out and identification of identical or equivalent structures in seemingly disparate contexts. For example, the description of algebraic, analytic, and geometric structures in the context of hyperbolic geometry, and the recognition that they are equivalent, are the source of important mathematical insights. The discovery of non-Euclidean geometry thus not only called for renewed attention to the axiomatic foundations, but it also—through the study of isometries of non-Euclidean geometries (Stillwell, 1996)—led to a renewed emphasis on the role of transformations and invariance in geometry.

The relation between transformations and invariance is old: One of the earliest examples is the "*Brouillon projet d'une atteinte aux événements des rencontres du cône avec un plan*" of Desargues in 1693, which showed that the properties of conic sections are (in modern terminology) projectively invariant, and therefore that a property proved true of one conic section is automatically true of all others (Hogendijk, 1991). But it was Felix Klein (1849–1925) who, 200 years later, developed the idea conceptually into a research program.

Klein became a full professor at Erlangen in 1872 and, following a tradition of all new professors, presented a working program to the Senate of the university: "A comparative review of recent researches in geometry" (Klein, 1893). The main thesis of this piece of work was to become one of the guiding principles in mathematics over the next few decades: Geometry, according to Klein, was best understood as the study of invariants under certain groups of transformations. He wrote: "The most important notion which is necessary in the following considerations is that of a group of spatial transformations. . . Let a manifold and a transformation group on it be given; one should investigate constructs related to the manifold with respect to properties which are not changed by the group" (p. 473). As an example, the triangle congruence theorems of classical Euclidean geometry (that is, Side-Angle-Side, Side-Side-Side, etc.) characterize the equivalence classes in a set of triangles with respect to the group of rigid motions in the plane.

In general, for any choice of a group acting on the plane we find a different set of invariants, and hence another type of geometry. For example, if we take as our group the set of all bijective maps in the plane, the only invariants are the cardinal numbers. In the same way, the group of rigid motions defines Euclidean geometry, the group of affine transformations define affine geometry, and the group of hyperbolic transformations define the hyperbolic plane.

Klein's Erlangen Program led to a complete reconceptualization of the nature of geometry; this in turn led to efforts to reorganize the school geometry curriculum along modern transformational principles. In France, transformations became one of three principles in the reform of the mathematics curriculum in 1902. (The other two were to no longer separate plane and solid geometry, and to emphasize the "experimental" nature of geometry; see Bkouche, 2009). According to Gispert (2014, p. 233), Poincaré posed the question "What is geometry?" only to answer his own question: "It is the study of a group. And of what group? Of that of the motions of solid objects" (Poincaré, 1904, pp. 18-19). This early inclusion of the transformational side of geometry in the French curriculum was preserved during a long period of stability in the mathematics curriculum (Laborde 1998). "Transformations of figures" were introduced in the French curriculum in 1925; in 1986, the programs still prescribed that students learn axial symmetry in grade 7 and point symmetry in grade 8, followed by a general study of translations and rotations (Bkouche, 2009). Other countries followed suit only much later than France, if at all. As Jones (2002, p. 128) points out with respect to the UK, it is difficult to determine how geometry was really taught over the course of the twentieth century due to lack of suitable contemporaneous data sources, but a review of the mathematics curriculum (by Willson, 1977, cited in Jones, 2002) set out "to show that the introduction of modern transformation geometry does not rule out the teaching of more traditional Euclidean type proofs, and indicates some of the many fruitful points of contact between the two areas," from which Jones (2002) concludes that "it is likely that geometry in the Euclidean fashion persisted in some places, while in others approaches based on transformations dominated" (p. 128).

The extent to which the transformation-based approach to geometry replaced the more traditional Euclidean synthetic approach is difficult to establish, but it is possible to point to a few anecdotal cases in which mathematicians advocated for (or in some cases wrote) transformation-based curricula. For example, in Russia, Andrei Kolmogorov in 1968 suggested that school mathematics be based on the study of transformations (Karp & Werner, 2011) and in the United States Usiskin and Coxford wrote a transformation-based textbook in 1972 (González and Herbst, 2006). Lately the Common Core State Standards for Mathematics in the US have prescribed the teaching of geometry using a transformation-based approach. This movement towards transformation-based geometry curricula surely responds to Dieudonné's proclamation at the Royaumont Seminar that "Euclid must go!" If geometry was to be understood in the modern era primarily as the study of the invariants of a space under a particular set of transformations, then what role was there for the naïve compass-and-straightedge constructions of Euclid? Mathematicians were pushing for updating the geometry curriculum to better reflect the ideas and ways of working of modern mathematicians, which had not only reduced attention to geometry (e.g., only three of the 23 problems Hilbert posed in 1900 were geometry ones) but also, through the influence of Bourbaki, had avoided the use of the visual register (Sinclair, 2008, pp. 46–49).

1.4.2. Geometry as an Opportunity for Students to Learn Logic

The second of the four modal arguments identified by González and Herbst (2006) was that geometry provided an opportunity for students to learn logical deduction and with this frame in mind we turn to another important change in the conceptualization of what geometry was and how it should be taught. Specifically, we refer to the movement toward a formalization of Euclidean (synthetic) geometry on a rigorous axiomatic foundation, one that addressed the logical gaps in the *Elements* and explicitly incorporated the tacit assumptions in Euclid's work.

The first major effort in this direction was Legendre's *Eléments de géometrie,* which first appeared in 1794. Well aware of the flaws in Euclid's *Elements,* Legendre set the specific aim in his preface to write a book that simplified and corrected Euclid. For this effort, he attempted to significantly reduce the set of axioms and to incorporate the (then-recent) developments in arithmetic and the Cartesian synthesis of algebra with geometry.

From a mathematical point of view, this approach was not without problems. For instance, Legendre made an uncritical use of arithmetic, neglecting to specify the number field he was using (see Hartshorne, 2000b, p. 462). In the course of trying to reduce the set of axioms in geometry, Legendre's *Eléments* is particularly known for its many purported proofs of Euclid's parallel postulate, none of which succeeded in his goal of demonstrating that it was reducible to Euclid's other four postulates.

The pedagogical influence of Legendre's textbook was considerable. It spread quickly in the academic world, with many translations. After nearly 2,000 years of the hegemony of Euclid's *Elements*, the radical idea of revising it for the sake of better studies for a wider audience fell on fertile ground following the French Revolution, the establishment of educational systems throughout Europe, and the rise of mathematics as a subject and a profession (Glas, 2002). Legendre's *Eléments* were translated into English beginning in 1819, and over the course of the next century went through more than 50 editions in the United States and Europe. Students' learning of logic through the study of geometry can be related to the activities students did and the logical qualities of the text of studies.

For the activities students did, texts such as Legendre's—and to a lesser extent Playfair's (1795)—set the stage for what Herbst (2002a) calls the "era of text": A period in the United States starting in approximately 1840 in which students' primary work in geometry was to study and master texts, with no expectation that they do any original work. The texts to be studied were often those by Playfair and Legendre, which, by virtue of having been written with attention to the logical development of ideas, were expected to induct students into logical thinking. From the 1850s to the 1910s the era of text gradually evolved, and students increasingly were expected to produce original proofs in the "era of originals," in which involving students in producing original proofs was expected to compensate for the shortcomings of possibly acquiring the text by memorization. Herbst (2002a) traces in the ensuing "era of exercise" the emergence of a set of norms for the writing of proofs that no longer were original problems but rather short exercises in deductive logic to prove rather inconsequential propositions, crystallizing in what came to be called the *two-column proof form*, and aimed at having every student succeed in learning logic.

In regard to the logical characteristics of the text of studies, Legendre's rewriting of Euclid was a step forward in the logical cleansing of Euclid's *Elements*. Up to the beginning of the twentieth century, Euclid's influence on schools was more significant and lasted longer in some countries than in others as documented in Stamper's (1909) study on the history of teaching geometry. In England, Jackson (1924, pp. 35–37) mentions a report of the Schools Inquiry commission in 1868, which named the following causes of difficulties with the teaching of Euclid's *Elements*: The lack of an introductory course, the ban on hypothetical constructions, the treatment of parallels, and the treatment of incommensurable magnitude (in book V). In *Euclid and His Modern Rivals* (first published 1879) the mathematician Charles Dodgson—better known, under his pen name Lewis Carroll, as the author of *Alice in Wonderland* and *Through the Looking-Glass*—critiqued 13 contemporary textbooks and argued for the continued use of Euclid's *Elements* as the main school geometry textbook. With respect to the Schools Inquiry Commission report, Dodgson wrote: "The Schools' Inquiry Commission has raised the question whether Euclid be, as many suppose, the best elementary treatise on geometry, or whether it be a mockery, delusion, snare, hindrance, pitfall, shoal, shallow, and snake in the grass" (Dodgson, 1879, p. 246).

In German schools, Euclid's *Elements* were sometimes used as texts, but only selectively and "with ocular demonstrations" (Stamper, 1909, p. 64; Barbin & Menghini, 2014). In Italy, Euclid was first introduced as a nationwide textbook in schools in 1867 before reforms detached education from his work (Giacardi, 2006). Menghini (2006) identifies Luigi Cremona as a key agent in bringing Euclid's text to the Italian schools. Interestingly, both Italian and German mathematicians prominently participated in the development of geometry in research, but it took a very long to bring this innovation to the secondary schools in these countries.

The impact of the mathematical theories of Hilbert and Klein on school mathematics was of an indirect kind. By repairing Euclid's trouble spots, Hilbert breathed new life into the synthetic approach to geometry and extended the usability of the *Elements* at school for a considerable time, as Freudenthal (1971, p. 433) claims: "Without doubt Hilbert's *Grundlagen der Geometrie* have contributed to lengthen the life of Euclid's methods for half a century, whereas Klein's algebraic approach was too little geometry to be interpretable in school mathematics." However, Freudenthal also notes that, regardless of its effect at reinvigorating interest in Euclidean geometry, very little of Hilbert's rigorous axiomatization found its way into actual classroom use: "At their best Hilbert's and Klein's approaches were teacher's [sic] background knowledge which hardly influenced school instruction" (Freudenthal 1971, p. 433).

The "New Math" movement set out to change all of that—to realign school mathematics along lines comparable to the radical changes that were characteristic of the cutting edge of mathematical research. The New Math had its roots in the 1930s, in a movement among (at first mainly French) mathematicians with the ambitious goal of rebuilding all of mathematics on a rigorous axiomatic foundation, much in the spirit of Hilbert's Program. These mathematicians—notably including Henri Cartan, Claude Chevalley, and André Weil—formed an anonymous collective, publishing their work under the pseudonym Nicolas Bourbaki. Over the course of the next half-century, Bourbaki produced an 11-volume comprehensive treatise on all of modern mathematics, the *Eléments des Mathématiques*. Bourbaki prized precise technical notation and scrupulously avoided the use of visual or intuitive methods; one footnote in Bourbaki (1950, p. 227) indicated: "Like all intuitions, this one also is frequently wrong."

The modernist confidence of the Bourbaki enterprise had a far-reaching influence on mathematicians at European and North American universities. Then, in the late 1950s, the shock in the West over the Soviet Union's successful Sputnik satellite launch catalyzed an international effort to reform the mathematics curriculum. The perceived urgency of the education reform project combined with the spirit of positivism of Bourbaki, giving rise to the "New Math". Guided by prominent mathematicians of the day—in the United States, particularly under the auspices of the School Mathematics Study Group (SMSG)—the new curricula were characterized by an unprecedented level of mathematical rigor, with set theory, functions, and proofs playing a crucial role at all grade levels.

It was against this backdrop that Jean Dieudonné, a member of the Bourbaki group, delivered his famous remarks at the Royaumont Seminar, calling for an end to the traditional privileged role of Euclid in the schools. Dieudonné considered the *Elements* a text of great historic interest, but felt that its content was too far removed from modern mathematics to be of value to students. Instead of focusing on the same contents as Euclid—mainly low-dimensional geometry without any links to algebra and analysis—Dieudonné proposed that the traditional geometry course be replaced with: a) Determinants of order 2 or 3, b) elementary calculus (function in one variable), c) construction of the graph of a function and of a curve given in parametric form (using derivatives), d) elementary properties of complex numbers, and e) polar co-ordinates (Dieudonné, 1959, p. 38). Dieudonné's remarks reverberated among mathematics educators, who continued to quote his controversial slogan "Euclid must go!" and debated its significance for decades to come (see e.g., Cornelius, 1973; Kapadia, 1980). Despite the influence of Dieudonné's call for reform, it took another decade, until 1969, for the "triumph of structure" to take hold in the French curriculum after a long period of stability since 1902 (Laborde, 1998).

In the United States, where the New Math briefly flourished in the 1960s and 1970s before being replaced by the so-called Back to Basics movement (Taylor, 1979), many of the topics that Dieudonné proposed should replace Euclid—for example, determinants of order 2 and 3, elementary properties of complex numbers, and polar coordinates—took root in the secondary advanced algebra or "Algebra 2" course, where they remain to this day. But the SMSG, led by topologist and mathematics education researcher Ed Begle, did not do away with Euclidean geometry, either as a topic of study or as a year-long course. The SMSG produced eight volumes devoted to geometry: A two-volume *Geometry* text and a two-volume *Geometry with Coordinates*, each with a corresponding two-volume Teachers' Commentary. In both texts, the SMSG covered a set of topics roughly corresponding to that of the traditional *Elements*-based geometry course, but with a consistent use of modern set-theoretic vocabulary and notation and with an axiomatic structure that prized economy and clarity over rigor for its own sake. In the Preface to the first volume of *Geometry*, the authors wrote:

> We began and ended our work with the conviction that the traditional content of Euclidean geometry amply deserves the prominent place which it now holds in high-school study; and we have made changes only when the need for them appeared to be compelling [. . .]
>
> It has been correctly pointed out that Euclid's postulates are not logically sufficient for geometry, and that the treatments based on them do not meet modern standards of rigor. They were improved and sharpened by Hilbert. But the foundations of geometry, in the sense of Hilbert, are not a part of elementary mathematics, and do not belong in the tenth-grade curriculum.
>
> *(Allen et al., 1965, p. 10)*

It is significant to note that although both SMSG geometry texts sought to modernize the presentation of the subject by fusing it with algebra and making use of the properties of the real numbers as a complete ordered field, neither text embraced the Erlangen Program-inspired reconstruction of geometry as a field whose basic objects were transformations (i.e., rigid motions) of the plane, and whose objects of study were the invariants under those transformations.

1.4.3. Geometry as an Opportunity for Students to Formalize Their Intuitive Experiences in the World

The third of Gonzalez and Herbst's (2006) modal arguments is that geometry offers a "unique opportunity for students to apply the intuition of the geometric objects to describing the world" (p. 20). Geometry provides a means for students to interact with the experiential world—through observation, measurement, and manipulation—and to interpret those concrete physical actions through the lens of geometric abstraction.

Consider, for example, the use of geometric tools, such as compass and straightedge, in the construction of geometric figures. The straightedge and compass are among the oldest examples of technology for geometry. Starting with a finite configuration of points, lines, and circles, the compass and straightedge make it possible to generate new points, lines and circles, through the iteration of five basic constructions, namely:

- constructing the line through two existing points,
- constructing the circle through one point centered around another point,
- constructing the intersection of two given lines,
- constructing the intersection(s) of a line and a circle (if any),
- constructing the intersection(s) of two circles (if any).

The use of compass and straightedge as generative tools in geometry is central to Euclid. Indeed, the first three postulates in the *Elements*—i.e., to draw a finite line segment between any two given points, to extend a given line segment arbitrarily in either direction, and to draw a circle with a given center and radius—say little more than: You have a compass and a straightedge, and you know how to use them.

Much of the first two books of the *Elements* is dedicated to showing just how much can be done with those simple tools. With compass and straightedge, one can bisect any angle (I. 9), bisect any segment (I. 10), construct perpendiculars and parallels (I. 11, 12, and 31), and produce a square equal in area to a given polygon (II. 14).

These accomplishments lead one naturally to seek after the limitations of the tools: Is there anything that *can't* be done with a compass and straightedge? This problem, one of the central questions of mathematics since antiquity, was definitively answered in the early nineteenth century using the methods of Galois theory in modern abstract algebra. In particular, Galois theory shows that the classical problems of doubling a cube, trisecting an arbitrary angle, and squaring a circle cannot be solved by the use of compass and straightedge alone.

In schools, constructions with a straightedge and compass have a long tradition as typical tasks. As noted previously, these constructions play an essential role in the development of synthetic geometry. From a cognitive point of view, construction tasks combine tactile manipulation with processes of visualization and reasoning (Duval, 1998). Roberts (2014) described the history of tools and technology in mathematics education with special regards to geometry. We return to some of these themes below, in our discussion of dynamic geometry software (DGS).

In addition to being recognized for the influence of his geometry research program, Felix Klein is regarded as the founder of mathematics education as a field of study at German universities. Klein called for a reform of mathematics education: In his opinion, the geometry courses suffered from the "burden of historical tradition" (Klein, 1925, p. 226), and he questioned the use of Euclid's *Elements* in schools.

Klein called for "functional thinking" as an organizing principle for mathematics (Krüger, 2000), arguing that it plays a fundamental role "to describe the life-world phenomena with metric predictors and represent dependencies between the so-quantified variables by means of mathematical functions." After the abolishment of biology from German schools by an intervention of the church at the end of the nineteenth century, Klein organized a commission of experts from medicine and science and worked out a vision for the sciences at school. He convinced the members of the commission to involve mathematics in the concepts for school reform. In the decisions of Meran in 1905, school mathematics was charged with two tasks: To "strengthen the spatial sense," and to "educate for functional thinking" (see Gutzmer, 1905, p. 61; also Allmendinger, 2013). The commission suggested that mathematics be implemented as a major at school with two programs, "Arithmetics" and "Theory of space" (*Raumlehre*), that were united in the final year into a single program. The topics to be covered in the "Theory of space" were as shown in Figure 1.1 below.

It took about a quarter of a century to see some consequences in curricula—not only in Germany. Several books were influenced by the shift, such as those by Behrendsen and Götting (1915), Treutlein (1911), or *Geometry and the Imagination* by Hilbert and Cohn-Vossen (1999), which first appeared in 1932. The long-term effect of the concept of functional thinking differs from its original intentions (Krüger, 2000).

This emphasis on using geometry to cultivate and refine students' intuition about space ran directly counter to the strict logical formalism of Hilbert's *Grundlagen*, which rejected ontological commitments and scrupulously avoided all arguments that relied on intuition. Charting a middle course between these two extremes in the United States were Birkhoff and Beatley, who wrote *Basic Geometry* (1941) for high school students. As has been noted above, Birkhoff and Beatley (1943/2000) (whose approach was later adopted by the SMSG in the New Math reforms of the 1960s) reduced the abstraction of Euclid's system by incorporating the properties of real numbers, via the measurement of segments and angles, directly into the

Grade	*Program in "Theory of space"* (Raumlehre)
10	Similarity theory by using similarity positions. Proportions at the circle. Calculations of approximate values for circumference and area of the circle by approximations with polygons. Intensive discussion of mutual dependence of proportions of sides and angles at triangles, in particular right triangles. Working out and verifying tables for this dependence (in preparation of trigonometry), hands-on tasks afterwards (recordings at the measuring table).
11	Trigonometry to refer to constructive planimetry. Connections to hands-on tasks on triangle and quadrilateral measurement. Characterization of the mutual dependence between change of angles and of functions by formulae in goniometry; graphical illustration of this dependence. Treating suitable tasks on multiple way, constructively and with the addition of calculations. Discussion of harmonic relations and of newer geometry as completion of planimetry.
12	Stereometry taking into account the most important elements of the theory of projections. Exercises in stereometric drawing. Simple theorems in spherical trigonometry. Mathematical geography including the theory of map projections.
13[1]	Theory of conic section in both analytic and in synthetic treatment with applications to elements of astronomy. Repetitions of the whole school curriculum, possibly with the help of bigger tasks which have to be carried out with calculations and drawings.

FIGURE 1.1 Program of Meran in 1905 for "Theory of space" in the Gymnasium from grade 10 ("*Untersekunda*")[2]

axiomatic foundations of their text: In essence, they supplemented the traditional compass and straightedge postulates with new protractor and ruler postulates, which stipulated that it was possible to assign numerical measurements to any segment or angle. Thus, whereas Hilbert (1902) required a collection of four "axioms of order" to specify an undefined ternary relation "between", Birkhoff and Beatley (1941) could simply appeal to the already-known ordering of real numbers to define what it meant for one point to be between two others. (See Sinclair, 2008, pp. 52–53, for a more detailed study of the underlying axioms.) This approach to geometry was initially not very popular, but it was later adopted by the SMSG and in the influential work of Moise and Downs (1964) and it continues to be used in school textbooks in the United States.[3]

The 'ruler axiom' asserts that the distance between any two points may be associated with an interval of real numbers, which—in addition to simplifying many proofs considerably (Smith, 2011, p. 42)—formalizes in abstract terms the intuition behind the physical act of measurement. In the above-mentioned preface of the SMSG book, the authors annotate: "If we assume the real numbers, as in the Birkhoff treatment, then the handling of our postulates becomes a much easier task, and we need not face a cruel choice between mathematical accuracy and intelligibility." (Allen et al., 1965, p. 10). The ruler and protractor axioms mark a striking departure from the classical, measurement-free approach to geometry; Hartshorne (2000b), commenting on the difference, wrote "It seems to me that this use of the real numbers in the foundations of geometry is analysis, not geometry" (p. 461).

Klein, too, had endeavored to take the learners' perspective (Hefendehl-Hebeker, 2000), arguing for a generic approach to education that stressed first phenomena and only later theory. "In particular," wrote Klein (1925, p. 227), "this means for geometry: First you have to provide a link to hands-on visualizations. Logical elements can be considered at a later stage." Klein promoted the use of manipulatives, both in research and in teaching. Various European universities house collections of objects used in hands-on approaches for teaching geometry and for research and they attest to the influence that Klein's research programme had.

Especially earlier in students' development of geometric thinking, the use of manipulatives is widely considered an important tool. "Perhaps the most important single proposition that the educator can derive from Piaget's work, and its use in the classroom, is that children, especially young ones, learn best from concrete activities" (Ginsberg & Opper, 1969, p. 221). In primary school, the use of concrete objects has a long tradition and has been considered fruitful by many educators for a long time. The mathematician E.H. Moore suggested in an address to the American Mathematical Society in 1902, the potential of concrete material for geometry: "The drawing and the paper folding must lead on directly to systematic study of intuitional geometry" (Trafton & LeBlanc, 1971, p. 17). Dienes (1969) pled for systematic use of things—that is, physical objects—in the learning of mathematics. The examples for geometry in primary school range from the use of mirrors, to challenging partitions of the cube, to paper folding. John Dewey in the United States, Maria Montessori in Italy, and Friedrich Fröbel in Germany developed educational programs structured around the use of manipulatives at primary school.

For older students, manipulatives have a long tradition as well, but their use is far less widespread in secondary school. Felix Klein considered manipulatives as more than merely tools: "A model—be it concretely in front of you or just vividly imagined—is not just a tool for a certain geometry. It is the object of study itself" (Klein, 1872, p. 42). Drawing, according to Klein, was also important: "Isn't the task of drawing correctly equally dignified for a mathematician as that of calculating correctly?" (Klein, 1895, p. 540). (Drawing was also a prominent activity in many other curricula, for instance in France; see Laborde, 1998).

In the Meran plans, Klein's influence suggested commencing with observations of geometry. For example, in grade 6,[4] the Meran program described the propaedeutical theory of space as follows:

> Introduction to the basic spatial concepts, however, so that the space appears primarily as a source of planimetric relationships. Spatial extensions, surfaces, lines, points are initially described in the neighbourhood and confirmed at various bodies. Plane figures initially as parts of the bodies' boundaries, then as independent structure on which the notions of direction, angle, parallelism, symmetry are brought to understanding. Exercises in the use of ruler and compass, continuous drawing and measuring.[5]

Later, more theoretical considerations on figures were intended, as will be illustrated below.

Visual representations in general and manipulatives in particular were said to enable professional teachers to "provide alternative explanations for students who [are] unable to follow the initial instruction" (Brophy, 1991, p. 352). Visualizations and other methods to strengthen intuition were also thought useful to bridge gaps of knowledge, as van der Waerden (1967) suggests as a different role for axioms in teaching and learning processes: "even in school lessons in geometry, it seems reasonable to me, to accept expressly or tacitly and not to prove all that is intuitively clear" (p. 2).

Geometry curricula structured around visualization, manipulatives, and other experiential forms of learning have one characteristic in common: A de-emphasis (or outright elimination) of the axiomatic as a unifying principle in geometry. Freudenthal (1971, p. 414) suggested that it was reasonable and appropriate to replace the "global organization" of an axiomatic system with "local organization"—allowing for traditional deduction to exist even in the absence of a comprehensive axiomatization of the theory. In making this proposal, Freudenthal was following in an earlier tradition. At about the same time as Legendre's publication of *Eléments de géométrie*, his countryman (and, to some extent, his rival) Sylvestre Lacroix had published his own identically-titled work, *Eléments de Géométrie* (1799). Lacroix's work broke with the classical exposition of geometry, in that the proofs were quite eclectic and aimed at connecting different ideas (Barbin & Menghini, 2014). It was shocking for many that the unifying power of an axiomatic system seemed not to interest the author. The work was produced in and for the "grandes écoles" and served as an important textbook example for engineering students.

Lacroix's approach to geometry built on the ideas of his teacher, Gaspard Monge. Lacroix introduced to mathematics the phrase "analytical geometry", writing that

> In carefully avoiding all geometric constructions I would have the reader realize that there exists a way of looking at geometry which one might call *analytic geometry* and which consists in deducing the properties of extension on the smallest number of principles by purely analytic methods, as Lagrange has done it in mechanics with regard to the properties of equilibrium and movement.
>
> *(Lacroix, 1797, p. xxxvii; in the translation of Boyer, 1956, p. 211)*

The *Eléments* were just one of Lacroix's many published textbooks, all of which were closely connected to teaching, due to the fact that Lacroix wrote for the preparation of his lectures (O'Connor & Robertson, 2004). However, despite the obvious value of algebra as a tool for studying geometry, in terms of education Lacroix felt that algebra and geometry "should be treated separately, as far apart as they can be, and that the results in each should serve for mutual clarification, corresponding, so to speak, to the text of a book and its translation" (Lacroix, 1797).

The movement away from a global axiomatic scheme found support later in the nineteenth century from Schopenhauer (1864), who described Euclid's proof of the Pythagorean Theorem as a "mousetrap" (Costanzo, 2008, p. 209) and argued that while "according to Euclid's treatment [. . .] the axioms are the only undemonstrable premises, and all demonstrations are successively subordinated to them" (Schopenhauer, cited in Emch, 1914, p. 525), this was not an essential characteristic of geometry; rather, he argued,

> every theorem begins with a new construction in space which in itself is independent of the preceding ones and which in reality can be recognized also in entire independence of them, in itself, by pure intuition of space, in which in reality also the most complicated construction is immediately as evident as the axiom itself.
>
> *(ibid)*

We see, then, that throughout the nineteenth and twentieth centuries, two simultaneous and opposite movements coexisted in the discourse of geometry: One (represented by Legendre, Playfair, and eventually Hilbert) focused on improving the rigor and consistency of Euclid's axiomatic system, and another (represented by Lacroix, Perry, and Birkhoff) focused on de-emphasizing axiomatics by grounding geometry in a foundation of experience, intuition, and other branches of mathematics.

1.4.4. Geometry as an Opportunity for Students to Learn Practical Applications

The fourth and final of the modal arguments for the study of geometry in high school identified by González and Herbst (2006) is a utilitarian one: Studying geometry teaches students skills that are useful in the workplace and other fields of application. Euclid's *Elements* were for a long time the most important scientific piece of work of all (Scriba & Schreiber, 2015). This was also considered important for craftsmen, as the following episode from Augsburg in the sixteenth century may illustrate: In this town, the production of lutes was booming and there was a lack of geometrical knowledge among the craftsmen. The town gave a scholarship over several years to the poor, but bright Holmann Holtzmann's (later known as Holmann Xylander) scholarship was tied to one condition: After his studies he was expected to translate the *Elements* into German. In 1562 he completed the translation of the first six books. In his translation he omitted several proofs, replacing some of them with exemplary calculations. In his preface, he commented that proofs "might be hard to understand for non-savants, and a simple German enthusiast of these arts is happy if he understands the matter even though he does not know the demonstration" (authors' translation from Schreiber and Brentjes, 1987, pp. 114–115).

This perspective may have been strongest in Protestant societies, where a more "practical" point of view of geometry was taken (Barbin & Menghini, 2014, p. 475). In 1871, the Association for the Improvement of Geometrical

Teaching (AIGT[6]) was established in Great Britain with the explicit aim to develop a replacement for Euclid, a goal which Dodgson ridiculed in his satire *Euclid and his Modern Rivals.*

The overall effect on schools of these debates at the end of the nineteenth century was not homogeneous, with both progressive and classical approaches finding support (Howson, 1982; Stamper, 1909). This mixed environment was similar in most countries; however, during the second half of the nineteenth century, schools in Britain and in the United States were under increased pressure to have their students study the *Elements* because of the importance of Euclid for passing university entrance exams (Brock, 1975, p. 22). A detailed account of the role of Euclid's system of axioms in textbooks for British schools in the early twentieth century, by Fujita (2001), shows that the logical order of Euclid's *Elements* was altered in order to gradually increase the level of sophistication of the theorems.

Applications of geometry were widespread internationally in the teaching and learning of mathematics in the nineteenth century. In the Collège de Québec, "mixed" mathematical topics, in addition to the pure mathematical areas, included "measurements of length, area, and volume, mechanics, hydrostatics, and spherical astronomy" (D'Ambrosio et al., 2014, p. 181). Another programme of "mixed mathematics" by John Davidson (1832) was used in Scottish secondary schools. The original, full title of Davidson's textbook shows that geometric contents were linked to the hands-on activities of craftsmen: "*A system of practical mathematics containing geometrical problems, plane trigonometry: The mensuration of heights and distances, surfaces and solids, conic sections, specific gravity, ships, cattle and hay, land and wood and artificers' works, gauging, gunnery, spherical trigonometry, geographical, geodesical and astronomical problems, and navigation.*" The opposing points of view of a deductive approach to geometry and a more experimental one are found in many countries; England was mentioned as an example earlier. Howson and Rogers (2014, p. 261) report that alternatives to Euclid were "written by practitioners rather than university fellows".

In the early twentieth century, the Meran plans were influential in Italy (Giacardi & Scoth, 2014, p. 218) and Germany (Schubring, 2014) for the introduction of calculus and functional thinking in secondary education but did not succeed in geometry. In contrast, at the same time in France, Poincaré's above-mentioned concept of transformation geometry was motivated by a more experimental approach. He regarded "the consideration of motions of solid bodies" as "the true source of geometry" (Poincaré, 1904, pp. 18–19, cited in Gispert, 2014).

In the North and South American colonies, where the school systems' curricula were first copied from those of the colonial masters, the drive towards emancipation and independence partially manifested as a shift towards more applied approaches. The arguments for such changes varied considerably. The need for urbanization in Mexico led to the concept of "subterraneous geometry" in colleges at the end of the eighteenth century (D'Ambrosio et al., 2014, p. 194) to develop background knowledge for engineering in the age of industrialization (Catalã, 1994, cited in D'Ambrosio et al., 2014). In Chile, a reform in 1931 introduced an "intuitive geometry course before the formal course on deductive geometry" (de Carvalho, 2014, p. 352). After consolidating independence in India, teaching and learning were discussed intensively; different approaches were tried and links to other fields were established (Kapur, 1990, p. 48, cited in Dauben et al., 2014) before the Association of Mathematics Teachers of India ultimately decided that "features of the new mathematics such as the emphasis on concepts should be incorporated into mathematics teaching; discovery methods and transformation geometry needed to be retained, without fetishizing the use of set theory or the axiomatic approach to geometry" (Dauben et al., 2014, p. 383). This synthesis was to build upon what Kapur (1990) called "relevant mathematics."

The independence of the Southeast Asian states was accompanied by a shift to applications within geometry, a subject that had previously been entirely devoted to exercises in proving. The Cultural Revolution in China rejected Western intellectualism and excised from the curriculum anything that did not appear to have an immediate practical use (Cai, 2002, cited in Dauben et al., 2014, p. 372), a radical reform captured by the slogan "Knock down the house of Euclid!" (Wang Yuan, 1999, p. 267, cited in Dauben et al., 2014).

Let us recall the influence of Bourbaki, which started in the 1940s, before skipping a couple of decades. The ideas of Bourbaki were in line with Hilbert's *Foundations of Geometry*, which began with the statement of its guiding philosophy: "Geometry, like arithmetic, requires for its logical development only a small number of simple, fundamental principles" (Hilbert, 1902, p. 1). Barbin and Menghini (2014) give a detailed account of the consequences of Bourbakism for geometry, in particular its impact on the mathematics education community. Freudenthal (1966) describes this transition as a radical break from a prior tradition, in which it was taken for granted that "geometry deals with real space—Pasch, Enriques, Veronese, Pieri, Klein stressed this" (p. 618), to a new paradigm in which

> the bond with reality is cut. Geometry has become pure mathematics. The question of whether and how to apply it to reality is the same in geometry as it is in other branches of mathematics. Axioms are not evident truths. They are not truths at all in the usual sense.
>
> *(Freudenthal, 1957, cited and translated by Bos, 1993, p. 56)*

In later writings, Freudenthal described geometry as "one of the best opportunities which exist to learn how to mathematize reality" (Freudenthal, 1973, p. 407) and characterized the future of geometry's role in the modelling of physical space as "endangered" (Freudenthal, 1971, p. 435).

Motivated at least in part by such considerations, resistance against the New Math grew in the 1970s. One major argument against the Bourbakism of the New Math was the growing expectation that the study of mathematics in school should enable students to apply mathematics in other sciences. In geometry, the reform pedagogy, with its emphasis on experiential learning and mathematical empiricism, was rediscovered. A prominent example was the "Realistic Mathematics Education" movement, which started with Freudenthal and prevented the New

Math from entering Dutch schools (van den Heuvel-Panhuizen, 2000, p. 3). Its main feature was to inspire and train both horizontal mathematization—i.e., moving between the "world of life" and the "world of symbols"—and vertical mathematization within the "world of symbols" (Treffers, 1978, pp. 78–79). "Realistic geometry" (Gravemeijer, 1990, p. 78) opposed the purely deductive approach to geometry and concentrated in primary and lower secondary education on six activities: Sighting and projecting; orienting and locating; spatial reasoning; transforming; constructing and drawing; and measuring and calculating.

In the PISA study (OECD, 1999), the discussion of the "mathematical literacy" domain directly invokes Freudenthal's claim that "Our mathematical concepts, structures and ideas have been invented as tools to organize the phenomena of the physical, social and mental world" (Freudenthal 1973, p. 41). Reflecting this perspective, the 2003 PISA framework included *Space and Shape* as one of four "overarching ideas," along with *Quantity*, *Change and Relationships*, and *Uncertainty* (OECD 2003, p. 35). Figure 1.2 shows the relative performance of our example countries in these four categories.

In *Space and Shape*, the Japanese and Swiss students performed significantly better than in the other areas, whereas the Dutch students were significantly weaker. These findings can be difficult to interpret, however. As an example, consider different countries' performance on the PISA 2003 "Carpenter" task, which asked students to determine whether a given quantity of lumber would be sufficient to build each of four different garden borders. Using both quantitative and qualitative methods, Ulfig (2013) investigated the role of geometry in problem solving processes with PISA tasks, and worked out students' problems with different conceptions of circumference while working at the carpenter task. Interestingly, the

	Mathematics	*Change and relationships*	*Space and shape*	*Quantity*	*Uncertainty and data*
Singapore	573	580	579	568	559
Japan	536	542	557	518	528
Switzerland	531	530	544	531	522
Netherlands	523	518	507	532	532
Canada	518	525	510	515	516
Poland	518	510	525	519	517
Germany	514	516	508	518	509
France	495	497	489	496	492
OECD	494	493	490	495	493

FIGURE 1.2 Total and section scores in PISA 2012 (OECD 2014)

Country	*No correct answer*	*One correct answer*	*Two correct answers*	*Three correct answers*	*All four correct*
Japan	1.3 %	13.7 %	13.0 %	31.9 %	36.9 %
Switzerland	1.7 %	24.1 %	17.7 %	30.0 %	22.4 %
Netherlands	0.8 %	22.9 %	19.3 %	31.1 %	25.3 %
Canada	1.1 %	26.0 %	19.9 %	31.3 %	19.2 %
Poland	1.7 %	24.3 %	19.4 %	29.7 %	21.7 %
Germany	1.9 %	23.9 %	18.5 %	28.6 %	23.6 %
France	1.7 %	25.1 %	17.0 %	30.7 %	18.3 %

FIGURE 1.3 Percentages of correct answers of the "Carpenter" tasks[7]

Dutch students performed better and the Swiss students worse than their performance in "*Space and Shape*" would otherwise suggest (see Figure 1.3).

As with the "Realistic Mathematics Education" movement, the "Back to Basics" movement of the 1970s in the United States (Taylor, 1979) marked a turn away from the formalism of the Bourbaki-inspired New Math and toward a re-discovery of the pragmatic, utilitarian approaches that had briefly flourished a century earlier. This movement found support in the growing expectation that school mathematics was to take a leading role in preparing students for applications in other sciences.

1.5. Recent Developments

This section discusses recent curricular developments in geometry, with particular emphasis on the effects of the availability of computers in classrooms. We take as organizing principles seven influential areas identified by Galuzzi and Neubrand (1998): (1) the impact of the computer, (2) changes in the development of mathematics itself, (3) greater emphasis on applications and modelling, (4) the "fundamental ideas" discussion, (5) constructivist ideas on learning, (6) the focus on mathematics as a human activity, and (7) geometry seen by students as an empirical theory. To avoid repeating earlier sections and to broaden the international perspective, we consider these seven organizing principles by comparing the curricula of different countries. In what follows, we focus primarily on the intended curriculum, that is the curriculum that is contained within national curriculum documentation (McKnight, 1987).

Hoyles et al. (2002) conducted a comparative study of geometry curricula on the basis of national policy documents from eight countries, which were compared to the curriculum in England. In the following sections we attend to the curricular themes identified by Galuzzi and Neubrand (1998) as we compare recent updates to those curricula. Figure 1.4 shows the national regions and curricula that we investigated.[8]

Country	Region (in case there is no national curriculum)	Latest year of curriculum in Hoyles, et al. (2002)	Latest curriculum version considered in this survey	Latest considered documents
Great Britain	England	1999	2014	The statutory programmes of study and attainment targets for mathematics at key stages 1 to 4[9]
France		1998	2010	Les programmes du college[10] MATHÉMATIQUES - SCIENCES PHYSIQUES ET CHIMIQUES Classes préparatoires au certificat d'aptitude professionnelle[11] Mathématiques, classe de première
Germany	Baden-Württemberg	1991	2006	Bildungsplan Gymnasium für die Klassenstufen 6, 8, 10, Kursstufe[12]
Japan		2002	2013	National Japanese Curriculum in Mathematics in Primary Education,[13] National Japanese Curriculum in Mathematics in Lower Secondary Education[14] National Japanese Curriculum in Mathematics in Secondary Education[15]
The Netherlands		1993	2014	Syllabus Centraal Examen, 2018[16]
Canada	Ontario	1999	2005	Ontario Curriculum 1 to 8 in Mathematics[17] Ontario Curriculum 9 and 10 in Mathematics[18] Ontario Curriculum 11 and 12 in Mathematics[19]
Poland		2001	2009	Podstawa programowa kształcenia ogólnego dla gimnazjów[20]
Singapore		2001	2006	Secondary Mathematics Syllabuses[21]
Switzerland	Lucerne	1996	2002	Lehrplan für die Sekundarstufe I Mathematik Zentralschweiz[22]

FIGURE 1.4 Sources for the comparative survey on curricula

Certainly, the intended curriculum tells only one side of the story; the curriculum that teachers enact may tell quite a different one. Any claims about the place of geometry in the curriculum are, therefore, hard to verify in reality without extensive observational studies. The 2003 version of TIMSS (Mullis et al.,

2004) found that, at least according to teachers' self-reported data, geometry and measurement still have quite a solid position in the enacted curriculum, in terms of the amount of time dedicated to it. In the same study, however, there were indications that performance in geometry varies more than in other areas, particularly in secondary education. TIMSS 2003 examined two topics in the area of 'measurement': (1) attributes and units and (2) tools, techniques, and formulae. The following topics were considered in geometry: Lines and angles, two-dimensional and three-dimensional shapes, congruence and similarity, locations and spatial relationships, and symmetry and transformations. Observations of the nine countries reveal interesting differences in their curricular frameworks. For example, the topic of similarity has shrunk or even has vanished entirely from different core curricula (e.g., Germany, the Netherlands). This development had previously been observed by Hoyles et al. (2002).

1.5.1. The Impact of Computers

As we have seen in the previous sections, tools have always been important in geometry, with construction tasks lying at the core of Euclid's *Elements*. Beginning with the arrival of scientific calculators in the 1970s, electronic technology began to have an impact on mathematics classrooms; in geometry, the ability to easily compute trigonometric functions particularly made certain kinds of tasks much simpler. With these new tools, the character of mathematics as a whole began to shift more to applications and modeling and strengthened analytical methods (Bright et al., 1994). At the same time, some geometric methods became redundant; for example, traditional methods of determining or approximating certain values for trigonometric functions were rendered obsolete. As the graphing calculator gradually displaced the scientific calculator in the 1990s, ready-to-use visualizations also reduced the need for certain geometric considerations, while simultaneously making many more mathematical topics accessible (Drijvers and Doorman, 1996). By the end of the twentieth century some scholars were arguing that the availability of technology should lead to an entirely new curriculum based not on computation but on application:

> Should not contemporary curricula be different from curricula in previous eras when calculators were not readily available and there was a much greater emphasis on pencil-and-paper algorithms? Similar curriculum and assessment questions arise for the middle and senior schools, as a result of the availability of contemporary hardware and software (like, for example, CD-ROM, calculators, Logo, Cabri Geometry, and advanced mathematics software such as Derive and Mathematica).
>
> *(Clements & Ellerton, 1996, pp. 164–165)*

From the earliest stages, the personal computer has been used to enhance the teaching and learning of geometry. The Logo programming language, developed in the late 1970s, offered a wide range of educational applications via its *turtle graphics* implementation, in which a cursor or "turtle" drew on a screen as it moved in discrete increments and rotated its orientation. Clements et al. (2001) show that studies in the 1980s found the effects of these tools on mathematics "ambiguous" (p. 7): Some studies claimed the use of Logo improved student performance, while other results were not as clear to interpret. Recognizing the potential of Logo, Clements, et al. (2001) created the Logo Geometry Project, a systematic, research-based elementary-grades curriculum. Their program set out to use the turtle graphics implementation of Logo in furtherance of three curricular goals: (1) "achieving higher levels of geometric thinking,"(2) "learning major geometric concepts and skills," and (3) "developing power and positive beliefs in mathematical problem solving and reasoning" (Clements et al., 2001, pp. 14–15). The Logo Geometry program was divided into three strands, each of which contained several activities. The three strands and their activities were:

- *Paths* (procedures for paths, turns, path/command correspondence and debugging, turtle deliveries);
- *Shapes* (squares, rectangles, equilateral triangles, regular polygons, parallel lines and parallelograms, classification of quadrilaterals); and
- *Motions* (introduction to mirror symmetry, introduction to geometric motions, Logo motions, rockets, congruence and motions, symmetry and motions, motions as flips) (see Clements et al., 2001, pp. 15–21, for more detail).

Beginning in the early 1980s, one of the major developments in the use of computers for teaching geometry was the development of systems that eventually came to be called *dynamic geometry software* (DGS). Motivated by the need for a tool for realizing geometric constructions, what emerged over the course of the last few decades was a wide range of sophisticated software platforms—The Geometric Supposer,[23] Geometer's Sketchpad,[24] Cabri,[25] Cinderella,[26] and GeoGebra,[27] just to name a few—which enabled far more than just nice visualization tools for geometric objects. DGS systems typically include some combination of the following features:

- *Dragging mode*: Most DGS systems enable the user to start from a set of free or initial objects (e.g., points plotted anywhere) and construct other objects, using tools analogous to classical compass or straightedge. The user can drag or change the position of the initial objects and then the position, orientation, or size of the constructed objects are modified in real time. Dragging mode thus constitutes the "ability to remember a construction as a procedure that can then be executed on other objects chosen or constructed by the student" (Kaput, 1986, p. 189; see more on this in Chapter 3).
- *Quasi-independence*: Created figures are operationally defined in the software by the construction procedures and the specification of the primitive objects on which those were applied. While the screen shows a particular diagram that

results from a particular location choice of the initial objects, the figure the computer stores is independent of its dragging history (Laborde et al., 2006).

- *Special features*: Most DGS systems include the ability for a user to create and save macros so as to be able to repeat constructions on different sets of primitives; they also permit tracing points as their position continuously varies in dragging mode; these are features that do not exist in paper-and-pencil constructions (Sträßer, 2002).

Thus in DGS, the diagram represented on screen at any given moment is only a single representative of an entire class of equivalent diagrams that can be continuously varied. The "diagrams" constructed in DGS are (at least potentially) fundamentally different from the static diagrams constructed in classical geometry (González & Herbst, 2009). For an instrumental analysis of dynamic geometry software refer to Leung et al., (2000). Whether DGS functions—to borrow the language of Pea (1985)—as merely an *amplifier* of traditional constructions or as a fundamental *reorganizer* of the object of study depends not only on the use to which DGS is put in the classroom but also on the features of the individual DGS implementation. One of the creators of Cabri Geometry, Jean-Marie Laborde, held that DGS constructions were, in an essential way, different from compass-and-straightedge constructions, and that therefore the geometry studied in DGS was a fundamentally different object to classical Euclidean geometry. Laborde reports that at the beginning of the Cabri project, the programmers tried to "make the distance" between Euclid's axioms and Cabri "as small as possible (. . .) But we decided nevertheless to introduce [the expression 'Cabri Geometry'] to make definite the point that what comes from the screen is not Euclidean geometry, it's not projective geometry (. . .) It has to be different" (Laborde, as interviewed by Scher, 2000, p. 42). In contrast, The Geometer's Sketchpad in its initial design was not intended to have a dragging mode at all; rather, its originally intended purpose was for students to be able to create highly precise but *static* figures from Euclidean geometry (Scher, 2000, p. 44). It was not until Nicholas Jackiw, who designed and programmed Sketchpad, recognized the potential of the mouse-driven user interface (then commonplace only on the relatively new Macintosh computers) that the decision was made to make Sketchpad dynamic. Reflecting on the role of this dynamic geometry Goldenberg and Cuoco (1998) have also argued that it gives access to a different kind of mathematics, a phenomenon that Balacheff (1993) has characterized as an informatics transposition of mathematical knowledge.

Using qualitative research methods, Hölzl (1996) described learners' strategies that show the challenges of creating and enacting learning designs that support a careful construction of figural concepts. Novices, as Hölzl (1996) showed, tend to avoid deeper analysis of a geometric situation; rather than engage with powerful ideas, they use supposedly easier, classical tools; tools are sometimes used in an unreflective way; and, finally, the goal of a task is altered—mainly to achieve it more quickly and easily. Thus, for example, rather than construct a dynamic diagram according to the *a priori* logical relationships among the diagram's elements, and exploit the power of dragging mode to observe which properties vary and which remain invariant, novices might instead simply produce a static drawing that 'looks right'. The goal of DGS use in educational settings is to create situations that bypass this tendency by creating a "need for reasoning" (Hadas et al., 2000). According to Sträßer's (2001) analysis, in DGS the world of geometry is lived in differently, broader in scope, with a new and more flexible structure, and offers more heuristic strategies. In addition, the technology makes it possible to involve modern geometry in the classroom.

New technology is in general a challenge for mathematics education, because it can change the didactical milieu (Lagrange et al., 2003). The integration of technology is not necessarily a question primarily of the curriculum, but rather may call for the development of different teaching praxeologies, as seen in a DGE (dynamic geometry environment, or a computer environment built with DGS) for exponential growth (Artigue et al., 2014). For students, the different worlds on the screen and of mathematical representations may be difficult to coordinate and relate (Boyd & Rubin, 1996). Foletta's (1994) study with The Geometer's Sketchpad revealed problems when students regarded the constructions as pictures and not, as intended, as solutions to a posed mathematical problem.

Laborde (1998) offered to teachers three types of tasks to promote "the link between visual evidence and geometrical facts" (p. 114): The "use of transformations as tools for obtaining geometrical properties" (p. 116), the construction of DGS drawings with imposed trajectories, and, finally, "black-box" tasks, where parts of the constructions are deliberately left unknown to the students. Certainly, DGS has broadened the technological repertoire for designing lessons. The use of appropriate learning strategies for both traditional and technologically sound elements is, however, quite subtle (Johnston-Wilder & Mason, 2005, pp. 240–256).

In recent years, the connections between DGS and computer algebra systems (CAS) and other tools have become more important, both in software—as in the freely available GeoGebra—and by the implementation of DGS in calculators for students. Geometric algorithms are also more accessible via Scratch[28] and other programming tools. Other emerging technologies provide new opportunities and call for new approaches; for instance, 3-D printers can now reproduce mathematical collections in schools and might allow for the creation of custom-built manipulatives by teachers (Bartholdi et al., 2011).

Let us finally look at the role national curricula give to dynamic geometry software. Of the countries whose national curricula were reviewed for this chapter, only Poland and Singapore do not mention DGS at all. Several of the nine curricula mention DGS among the tools students should get acquainted with, often in quite general terms (the National Japanese Curricula, for example, say only that "It is necessary to appropriately use computers and so on, when necessary, in order to enrich pupils' sense of numbers, quantities and geometrical figures"). In most of these countries it is completely left to teachers to decide where and how this should be done. A notable exception is France, where, at the end of the first year of the

Collège (or junior high school, for students aged 11 to 15), the curriculum explicitly requires that students learn to "construct simple figures with the help of dynamic geometry software". However, other mentions of dynamic geometry software in the French curriculum are only suggestions, not directives. In French schoolbooks, the use of dynamic geometry in tasks has become frequent (Laborde, 2008).

The province of Ontario in Canada (which has a province-wide license for DGS; Nathalie Sinclair, personal communication, April 2016) refers to DGS as one of many tools that can be expected to be used by grade four, when students learn to "extend and create repeating patterns that result from reflections, through investigation using a variety of tools (e.g., pattern blocks, dynamic geometry software, dot paper)". Similar suggestions are made for other areas and for older students, as well. In the curriculum's list of "Sample problems" the suggestion is made to use DGS to promote geometric reasoning: "Use dynamic geometry software to show that parallelograms with the same height and the same base all have the same area." Even though dynamic geometry software is never specifically mandated, it is more often mentioned from grade 9 and, there, its use as a "black box" is extended, as in the case of trigonometric functions. The more frequently dynamic geometry software is mentioned, the more its use tends to be interdisciplinary.

Other curricula in Europe and North America mention dynamic geometry software, typically in the middle grades (specifically grades 7 or 8). Regional differences might be related to the presence of dynamic geometry in continuing education and professional development for teachers; for example, the software has a more prominent role in France, home of Cabri, or in the state of Florida in the USA, where the GeoGebra Academy has been established for a longer time. Although DGS implementation has been inconsistent across regional and national curricula, at the same time programming with Logo and similar tools has essentially vanished; in contrast, the use of physical manipulatives seemed to gain more attention at the beginning of the twenty-first century.

1.5.2. Changes Due to Developments in Mathematics

One might expect that new developments in mathematics would naturally have their effects on the curriculum; indeed, much of the New Math reforms was based on the premise that school mathematics needed to be radically reconstructed to bring it closer in line with contemporary mathematics, as envisioned by the Bourbaki group. However, in general, curricula often are largely unaffected by scientific innovation, as Cuban (1992) has documented. As we have seen, many countries largely ignored the Erlangen Program, resisting the shift to a transformation-based geometry curriculum and preserving the classical, construction-based geometry of Euclid. It may be worth noting, however, that dynamic geometry—because of its implementation of rigid motions—may be responsible for the incorporation of aspects of transformational geometry in all of the countries studied by Hoyles et al. (2002).

Most new developments in mathematics occur at a very high level of abstraction and specialization and therefore are arguably ill-suited for classroom implementation. But there are exceptions; one notable example is "fractals for the classroom" (Peitgen et al., 1992). The arrival of the personal computer in the 1980s enabled schools to generate and explore fractals, which have properties (for example, self-similarity and non-integer dimension) that challenge the boundaries of classical geometric constructions and transformations. However, in the latest versions of the nine national curricula we reviewed, no hint of fractal geometry was found.

Fractal geometry is not the only "new" kind of geometry that has aroused interest in both pure and in-school mathematics. Discrete geometry (for example "geoboard geometry" or "taxicab geometry") is another example. Detailed suggestions for integrating these geometries into existing curricula have been developed (see for example Kenney and Hirsch, 1991). Graph theory—the study of discrete networks defined by vertices and edges—allows for building up a restricted theoretical framework within which students can experience success in arguing and modeling. Algorithmic solutions to optimization tasks appear within modeling processes (Fey, 1984).

1.5.3. Greater Emphasis on Modeling and Applications

In 1973, the NCTM published its yearbook on "geometry in the mathematics curriculum" (Henderson, 1973). The book asked to what extent the one-year geometry courses at high school should be governed by Euclid's approach. Today, the yearbook stands as evidence of a critical moment in mathematics education in the United States, at the waning days of the New Math and the beginning of the "Back to Basics" movement as educators grappled with (in the language of González and Herbst 2006) "competing arguments" for why teach geometry. In Henderson (1973), the "applications" mentioned are mainly intra-mathematical applications of geometry. Only Forbes' (1973) contribution to the Yearbook names "real-world applications" as an objective for teaching an "elective program in geometry." But by the end of the twentieth century, mathematics was widely expected to involve applications and mathematical modeling. In the internationally comparative PISA tests, which focus on "mathematical literacy", there are clear differences between geometric tasks in a modeling framework and questions posed rather in the spirit of pure mathematics (Knoche, Lind, Blum, Cohors-Fresenborg, et al., 2002).

1.5.4. The "Fundamental Ideas" Discussion

With the exception of the regional curriculum of Ontario, Canada, most of the nine curricula reviewed in this chapter follow a spiral design, i.e., a design in which topics are revisited in successive years. In geometry, too, we find that certain units of geometry are found in almost every grade. When a new unit in geometry starts, the learners have an opportunity to recall and reinforce their memories of content taught the previous year. At the same time, this can lead to the geometric content of the curriculum being experienced by students as disjointed and disconnected.

In much the same way that the 2003 PISA exam conceived of mathematical literacy in terms of a collection of "overarching ideas" (see section 1.3 above), Schweiger (2006) has argued that the components of a mathematical curriculum can be given some structure by framing them in terms of *fundamental ideas*. Schweiger (2009, pp. 65–66) lists several different candidates for these "fundamental ideas", such as *symmetry, invariance, number, exhaustion, algorithm, functional thinking*, and so forth. Each of these fundamental ideas represents an independent axis along which competencies may be built up in a cumulative way.

Despite widespread agreement that these fundamental ideas are useful guidelines for developing curricula, there are nevertheless substantial disagreements among national curricula as to which topics are considered appropriate at what age. For example, in Japan certain geometric concepts are introduced much earlier than in most other countries. The concept of triangle congruence, for example, is introduced in grade 5 in Japan, whereas it does not appear for the first time until grade 7 in Great Britain, Germany, and Canada, and not until grade 8 in Switzerland and Singapore.

Geometric reasoning is present in the intended curricula of all the investigated countries; it remains unclear, however, to what extent "reasoning" is meant to include what might be called "proof" in geometry. Since several national curricula explicitly separate *processes* (like proving) from specific *content areas* (like geometry), it is no longer the case that secondary curricula consistently prescribe that proofs be an integral part of a geometry course. This does not mean that proving is no longer expected, but rather that teachers are increasingly left to their own discretion as to when and how to engage their students in proof, and whether this should be done in algebra, analysis, or geometry. In any case, most of the countries whose curriculum we have reviewed conclude the bulk of their geometry topics (although not analytic geometry) well before students take school exit exams or college entrance exams.

Interestingly, Japan and Poland—both widely celebrated for their successes in international school comparison studies—have quite ambitious geometry programs, with geometric reasoning and proving explicitly required in both countries. In Japan, the findings of the TIMSS video study (Hiebert et al., 2003) and textbook research (Fujita and Jones, 2014) underline ongoing efforts to push for problem solving and deductive reasoning.

1.5.5. Constructivist Ideas on Learning

The question of how knowledge is constructed has haunted many philosophers and been part of the historical debates on geometry. According to Richards (1988, pp. 25–35) the positions of rationalism and empiricism came into conflict in Victorian England. The dominant rationalism of Humanistic education was challenged by empiricists like Herschel and De Morgan, for whom geometry provided knowledge of the real world, and John Stuart Mill, who considered mathematics as a descriptive science, similar in nature to the sciences. Whewell sympathized with Kant's transcendental idealism and regarded geometry as *a priori* knowledge, which exists independently of experience.

Educators in Europe and North America have long pushed for reforms in the educational systems that would place students' own activity at the center of the learning experience. At the end of the nineteenth century, the ideas of John Dewey, Stanley Hall, Johann Friedrich Herbart, Johann Heinrich Pestalozzi, and Edward Lee Thorndike were particularly influential in promoting a student-centered approach to education. In the publications *How We Think* and *The Influence of Darwin on Philosophy and Other Essays in Contemporary Thought*, Dewey described how human knowledge is created as a process of evolution. In the specific context of teaching geometry, Dewey opposed the point of view that learning consists of acquisition or accumulation of knowledge (Dewey, 1903): Rather, he argued that learners should be given opportunities to construct knowledge for themselves out of the raw material of their own experiences. Dewey called for "material of reflective inquiry not as ready-made intellectual pablum to be accepted and swallowed as if it were something bought at a shop" (Dewey, 1933, p. 257).

As an educator, Dewey thus represented the flow of American progressive education, corresponding to a similar movement in Europe, led by Maria Montessori (1870–1952), Rudolf Steiner (1861–1925), and Celestin Freinet (1896–1966). Their impact on the education of children was much more immediately apparent at the primary level than it was in the secondary school, and will be taken up below in the section on manipulatives. One common thread connecting these progressive educators was the principle that students' learning should be activated by enabling them to carry out experiments. An earlier application of these ideas in geometry was manifest in Spencer's (1876) *Inventional Geometry* (Sinclair, 2008, p. 21), and the approach later found expression in the Perry Movement in Britain. Perry (1901) called for more time for elementary and "experimental" geometry (that is to say, physical measurement of objects, including estimation and approximation), as well as the introduction of trigonometry and 3-dimensional coordinate geometry, and sought to reduce abstraction, attention to axiomatic foundations, and the strict adherence to Euclid that was then predominant in Britain. Educational reforms led to further experimental approaches in the mathematics classrooms in many countries, for instance in France (Gispert, 2014, p. 234), in Germany's Weimar Republic (Schubring, 2014, p. 250), and in Japan (Yamamoto, 2006). The idea of generating new experimental curricula was perverted in Nazi Germany, in which the geometry program for girls was reduced to the study of objects in the household (Schubring, 2014, p. 251).

These reforms paved the way for the constructivist learning theory. If the goal of instruction in geometry is (at least in part) to support students in developing and formalizing their intuition about their experiences in the world, it is only natural that geometry instruction take note of psychological theories of learning about space, shape, and quantity. In subsequent chapters of this book, we devote more attention to some theories concerning the learning of geometry; here we briefly mention some of the main protagonists in that field because they played a role in the ongoing discourse about what teaching geometry should be about.

Almost from the beginning of psychology's emergence as a modern science, psychological aspects of child development were important considerations for the learning of mathematics, for instance those by the philosopher and psychologist Johann Friedrich Herbart (Greenstreet, 1894). Psychology delivered documented empirical research that could inform questions about mathematics education. Particularly relevant for geometry were studies of visual thinking. Louis L. Thurstone (1932), a US pioneer in the field of psychometrics, identified the "spatial sense" as one of seven independently measurable factors of intelligence. Later, he described three subfactors of the spatial sense: S(1) spatial relations, S(2) visualizations, and S(3) spatial orientation. Linn and Petersen (1985, 1986) distinguish between visualization, spatial perception, and mental rotation. Frostig (1963) studied visual perception and set up a training program for developing it.

The developmental psychologist Jean Piaget, who was influential in much of modern educational research, articulated the epistemology known today as constructivism. Piaget himself studied the construction of geometric ideas, in joint work with Inhelder and Szeminska (Piaget and Inhelder, 1956; Piaget et al., 1960). The main focus of this work, which has been revisited in more rigorous research designs over the ensuing decades, was children's development of spatial abilities. According to the Piagetians, children distinguish first topological, then basic projective and then basic Euclidean relations before developing multiple projective and, finally multiple Euclidean relations (zur Oeveste, 1987). Pierre van Hiele and Diana van Hiele-Geldof built on Piaget's work to develop their levels of geometric thinking (Fuys et al., 1984), as we shall see in Chapter 3. The van Hiele model became both an important research area and a source of much debate, with many consequences for curricular developments in geometry.

Constructivist learning theory holds that students learn not by passively listening, observing, or otherwise receiving information, but through a process of actively constructing their own knowledge, adapting their cognition to assimilate their experienced world. Curriculum designers have developed from that descriptive statement the design principle that contexts and activities need to be created that can activate students' thinking. The proposals by Johnston-Wilder and Mason (2005) are based on the premise that geometry tasks can play an important role in creating such contexts. Of course, hands-on tasks such as these are hardly new to the curriculum; Fujita and Jones (2003) studied the role of hands-on tasks in three textbooks written by Godfrey and Siddons during the years 1900-1920. Fujita and Jones note that all three texts are constructed in accordance with the principle that "practical and deductive geometry should be combined" (p. 57). Godfrey, in articles written during the period, argued that "There must be a good foundation of practical work, and recourse to practical and experimental illustration wherever this can be introduced naturally into the later theoretical course. Only in this way can the average boy develop what I will call the geometrical eye" (Godfrey, 1910, cited in Fujita and Jones, 2003, p. 54). Fujita and Jones (2003) conclude that "it is illuminating that innovative teachers 100 years ago pointed to the importance and roles of visual images in geometry and geometrical thinking" (p. 58).

Schoolbooks following a constructivist approach generally seek to interweave visual and mental operations; examples can be found in the Swiss curriculum and the book series *Mathbuch*.[29] Jones and Mooney (2003) elaborate on the use of mental geometry problems. An example of mental geometry questions is "What shadows can a cube cast" (Jones & Mooney, 2003, p. 10); other examples can be found in the online course by Conway et al. (1991). In such problems, the 'drill and practice' of algorithms is neither intended nor possible. Instead, solving mental geometry problems requires the abilities to imagine geometrical structures, vary sizes and shapes, combine them, rotate them spatially, transform them, and apply knowledge to them. The suggestion is that practice with mental geometry can help students pass from the "visual level" to the "abstract level" of van Hiele (1964). Because they require no materials or manipulatives, mental geometry problems can be deployed quickly, for example at the beginning or end of each lesson (possibly in conjunction with mental arithmetic practice).

1.5.6. Mathematics as a Human Activity

Over the twentieth century, there has been a growing appreciation for curricula that represents mathematics as a human activity, looking beyond the facts and procedures that it has produced to how mathematicians do mathematics. As far as geometry is concerned, Euclid's *Elements* has a distinctive expository style, one that has, over the centuries, come to be standard for advanced mathematics textbooks. Lakatos (1976) described this as a "deductivist style", which "starts with a painstakingly stated list of axioms, lemmas and/or definitions. The axioms and definitions frequently look artificial and mystifyingly complicated. One is never told how these complications arose. The list of axioms and definitions is followed by the carefully worded theorems. These are loaded with heavy-going conditions; it seems impossible that anyone should ever have guessed them. The theorem is followed by the proof" (Lakatos, 1976, p. 142). What the deductivist style conceals, notes Lakatos, is the process by which the theorem and its proof were originally discovered, which may have been very different from the final form in which it is presented.

Proofs are hard to achieve, and may be even harder to teach (Cirillo, 2013). The NCTM standards on geometry call for reasoning across all grade levels, but acknowledge that mathematical proof should be reserved for older students:

> At all levels, students should learn to formulate convincing explanations for their conjectures and solutions. Eventually, they should be able to describe, represent, and investigate relationships within a geometric system and to express and justify them in logical chains. They should also be able to understand the role of definitions, axioms, and theorems and be able to construct their own proofs.
>
> *(NCTM, 2000, Chapter 3, Geometry, p. 42)*

The original NCTM (1989) Curriculum Standards had called for a de-emphasis on the two-column proof form that was (and to a great extent remains) common in high school geometry in the US (Herbst, 2002a). Horgan (1993) noted that a then-recent seminar organized at the Mathematics Science Research Institute in Berkeley was organized around the topic "Are proofs in high school geometry obsolete?" Yet two-column proofs have continued to be taught to students, alternative forms of proof are less common, and the recommendation to de-emphasize two-column proofs was not found in later versions of the NCTM Standards (Herbst et al., 2009; NCTM, 2000).

There is a need for students to acknowledge the human activity that precedes the formalization of a written proof; but this requires a broader understanding of what counts as proof. Reid and Knipping (2010) articulate a wide variety of different kinds of argument structures that may be labelled as "proof," ranging from simple arguments to more detailed reasoning, and they examine their roles in classrooms (see also de Villiers, 1990, 2003). Cabassut (2009), however, notes that most textbooks do not prepare students for this variety of different proving activities. In fact, Hanna and de Bruyn (1999, p. 180) claim that "only in the topic of geometry do the textbooks do a reasonable job of providing opportunities to learn proofs".

But whether geometry textbooks play this role adequately is a contentious issue. The tradition of the two-column proof, studied in detail by Herbst (2002a), is often critiqued as a formal "ritual proof scheme" (Harel and Sowder, 1998) that does not help develop heuristic strategies for problem solving; however, Weiss et al. (2009) argue that the norms associated with the two-column proof form can also support the process of discovering an argument, depending on how the teacher uses those norms in his or her management of the class's mathematical practice.

1.5.7. Geometry as Empirical Theory

Particular attention has been paid to whether dynamic geometry helps to support the teaching and learning of proof, or whether it undermines the need for it by providing seemingly conclusive empirical evidence for the truth of a claim. Hanna (2000) argues in favor of embedding proving activities in the context of dynamic geometry. De Villiers (2004) suggests that space be made in curricula for the role and function of proof with DGS by creating opportunities for students to experience the role of proof as verification (assuring the truth of a statement), explanation (providing insight into why it is true), systematization (helping organize results into a mathematical system), discovery, communication, and intellectual challenge.

Graphing calculators and computers have brought a rich repertoire of tools for exploration and experiment. The nine national or regional curricula reviewed for this chapter all consider such explorations and experiments valuable; some of them place exploring problems as a priority as important as reasoning. The 2007 curriculum for secondary education in Singapore, for instance, directly addresses the duality of procedural skills and underlying mathematical principles: "Although students should become competent in the various mathematical skills, over-emphasizing procedural skills without understanding the underlying mathematical principles should be avoided. Skill proficiencies include the ability to use technology confidently, where appropriate, for exploration and problem solving" (Curriculum Planning and Development Division, 2006, p. 3).

Struve (1990) conceptualizes the processes of learners who build up their own theory by their experiences with and insights into the real world; such a theory is normally called "empirical theory" (Sneed, 1971). There are many suggestions of how to use dynamic geometry environments for exploration, for instance to (re) discover theorems (Wares, 2010), and for research on the interplay between inductive investigations and deductive reasonings, for instance Chazan and Yerushalmy (1998) and Sinclair and Robutti (2012).

1.6. Conclusion

This chapter has collected a range of issues that have been discussed over centuries of teaching and learning geometry. Some of them appear in antipodal pairs again and again, in particular the dialectic between practical and intuitive approaches to geometry on the one hand, and that of deductive and rational approaches on the other. Similarly, we have encountered arguments for visions of geometry that are *applied* and others that are *purely mathematical.* Some stress that geometry is *intrinsically* important, whereas others stress its enormous *interdisciplinary* potential—both in other branches of mathematics and in other disciplines. The possible uses of the computer are another, more recent development, one that each curriculum must attend to and every teacher has to negotiate in everyday coursework.

Klein's research programme in mathematics, which sought to establish a new foundation for how geometry should be understood, emphasized how figures change under certain classes of transformations. But in his work on the mathematics curriculum, Klein looked beyond purely mathematical considerations and regarded four principles as important for the teaching and learning of mathematics (Hefendehl-Hebeker, 2000): (1) taking psychological aspects of learning into consideration, (2) providing a stronger link to the phenomena and applications in the sciences and engineering, (3) teaching spatial and plane geometry simultaneously and, if appropriate, also arithmetic and geometry, and (4) enhancing functional thinking which, in geometry, means to give space to the "new spirit of geometry:" "the manoeuvrability of every figure, with the help of which the general character of a geometric construct becomes viable every time beyond the special case" (Klein, 1925, p. 228). The notion of figure was still essential in Klein's geometry.

The same is true in regard to work on students' thinking and learning of geometry. In Chapter 3 we discuss how different levels of understanding can be inferred from observing how students relate to geometric figures, coordinating figural and conceptual aspects of a geometric configuration. Fischbein (1993) argues that geometry, more than other mathematical disciplines, always involves a close coordination of percepts and concepts; indeed Fischbein introduces the notion of "figural concept" to represent this fusion of the visual with the abstract, involving mental entities which are neither pure concepts nor mere images. This uniquely visual character enables students to draw on their intuition and to engage in exploration, which can in turn lead to complex reasoning and proofs via a progressive mathematization (Jones, 2000). Dynamic geometry provides a new medium in which Klein's focus on the maneuverability of figures can be realized, potentially enabling students to build up figural concepts with the help of diagrams (Laborde 2005). In the following chapters, we take up the role of these figural concepts as an organizing feature of our analysis of the teaching and learning of geometry.

Notes

1 In the last grade, the distinction between "Arithmetics" and "Theory of space" is given up; the points are mentioned without specification of these areas.
2 Contained in the appendix of Klein (1907).
3 Some popular examples are: a) Bass, L., Charles, R., Johnson, A. and Kennedy, D. (2004). Prentice Hall Geometry. b) Larson, R., Boswell, L. and Stiff, L. (2004). McDougal Littell Geometry. Houghton Mifflin. c) Boyd, C., Cummins, J., Malloy, C., Carter, J. and Flores, A. (2005). Glencoe Geometry. McGraw Hill.
4 "Quinta" is translated here as grade 6.
5 Contained in the appendix of Klein (1907).
6 AIGT renamed to the Mathematical Association in 1894 (http://www.m-a.org.uk/)
7 Numbers for 0 correct answers do not include the case of no answer; this is why the values do not add up to 100 %.
8 Some countries organize schools on the level of regions or federal states. In these cases, the region picked by Hoyles et al. (2002) was selected again. The dates indicate the last year of publication of versions of curricula used for this survey.
9 https://www.gov.uk/government/publications/national-curriculum-in-england-mathematics-programmes-of-study
10 http://www.education.gouv.fr/cid81/les-programmes.html#Math%C3%A9matiques
11 http://www.education.gouv.fr/cid50637/mene0925433a.html
12 http://www.bildung-staerkt-menschen.de/service/downloads/Bildungsstandards/Gym/Gym_M_bs.pdf
13 http://www.mext.go.jp/component/a_menu/education/micro_detail/__icsFiles/afieldfile/2009/04/21/1261037_4.pdf
14 http://www.mext.go.jp/component/a_menu/education/micro_detail/__icsFiles/afieldfile/2011/04/11/1298356_4.pdf
15 http://www.mext.go.jp/component/a_menu/education/micro_detail/__icsFiles/afieldfile/2012/06/06/1282000_5.pdf
16 www.nvon.nl/sites/nvon.nl/files/wav2018syl.pdf
17 http://www.edu.gov.on.ca/eng/curriculum/elementary/math.html
18 http://www.edu.gov.on.ca/eng/curriculum/secondary/grade9.html
19 http://www.edu.gov.on.ca/eng/curriculum/secondary/grade11.html
20 http://www.bip.men.gov.pl/men_bip/akty_prawne/rozporzadzenie_20081223_zal_4.pdf
21 http://www.moe.gov.sg/education/secondary/normal/
22 http://alt.zebis.ch/Unterricht/Fach/Unterrichtsmaterialien/dateizeigen.php?id=4216&vz=rN2Kx6Eh2aW54t2fDydN5csma5nfUL&file=Lehrplan+Mathematik+Sek+I+LU+2002.pdf
23 See Yerushalmy and Houde (1986).
24 http://www.dynamicgeometry.com
25 www.cabri.com
26 http://www.cinderella.de/tiki-index.php?page_id=2
27 www.geogebra.org
28 https://scratch.mit.edu/
29 *Mathbuch 2 Begleitband* (2014). Zürich: Klett und Balmer. See www.mathbu.ch

2

GEOMETRIC FIGURES AND THEIR REPRESENTATIONS

2.1. Introduction

The notion of geometric figure, while defined loosely as well as restrictedly by Euclid ("14. A figure is that which is contained by any boundary or boundaries", Euclid, 1956, p. 153), is central to Euclid's *Elements*: Postulates enable possible figure constructions, and propositions assert properties of those figures or demonstrate that other figures can be constructed. Later expositions of geometric knowledge have varied in the extent to which they center on the notion of figure, with modern treatises either making figure a centerpiece as an application of set theory ("by figure we mean a set of points", Moise, 1974, p. 37) or making figure a derived notion within a more general consideration of geometry as the study of transformations of space onto itself (see for example Guggenheimer, 1967; see also Jones, 2002; Usiskin, 1974).

The notion of figure has also played a crucial role in scholarship on the teaching and learning of geometry (e.g., Duval, 1995; Fischbein, 1993). This chapter considers how scholarship from various disciplines has influenced our community's thinking about geometric figures. It brings perspectives from mathematics and mathematicians, from the history and philosophy of mathematics, from cognitive science and semiotics, from technology and from mathematics education proper. Influenced by those perspectives we elaborate on the role of the geometric figure in geometry teaching and learning.

We articulate some conceptions of figure related to its various representations in the context of making a curricular proposal on which to found research and development: To conceptualize the study of geometry in secondary schools as a process of coming to know figures as mathematical models of the experiential world. By this we mean that the study of geometry in secondary school could be seen as the development and use of conceptions of figure for students to make a particular kind of sense of their experiences with space and shape in the world: A sense-making akin to that of an applied mathematician creating a model of a real world phenomenon.

This assertion clearly is not meant as a statement of fact—we have not ascertained that such a position is articulated by a majority of curricular offerings in geometry. Nor do we intend it prescriptively or ideologically, as a statement of how we think curriculum ought to be designed. Rather, the assertion is meant as the proposition of a perspective from which to read existing research: If the study of geometry were to be conceived of in that way, what would the existing research have to say about it? The assertion is also intended as the proposal of a large design research project within which to ask further research questions: If the study of geometry could be conceived of in that way, what could it look like, what could it make possible as far as students' thinking and learning, what knowledge might it require of teachers, and what demands and tensions might it pose for teachers?

While in Chapter 1 we describe how geometry and its teaching and learning developed historically, the present chapter expands on one way of conceiving of the secondary geometry curriculum that can arguably help organize the existing research and promote further research and development. Thus, this chapter is, broadly speaking, a chapter about what there is to be learned in secondary school geometry. It plays on the theme that secondary geometry is the study of geometric figures and elaborates on that theme as it makes it more and more complex. We do not pretend that geometry is solely the study of figures but we do contend that a focus on geometric figures is at the core of any viable study of geometry in secondary school, and that any comprehensive reorganization of the subject matter (e.g., as a study of transformations of the plane or space) would require working out a transition from an earlier consideration of the geometric figure.

From a conceptualization as the study of geometric figures, the secondary geometry course can also take care of some of the other goals traditionally ascribed to the teaching and learning of geometry, including learning to master space and learning to craft proofs. We assert this, in particular, because students do not encounter geometry for the first time in secondary school. Rather, they come to secondary schools with knowledge of geometry that has been building up since they started to interact with the world through movement, observation, play, and talk, and through their primary education. Much of that interaction has been enabled by things and indexed by signs that relate, in various ways, to geometric figures.

Thus our first move is to argue that when students come to secondary geometry they already have some conceptions of figure even if they don't necessarily use the word *figure*. These conceptions of figure are ways of making sense of their activity at various levels of spatial organization. Those conceptions of figure are, at the very least, prior knowledge upon which new geometric experiences will be built. Furthermore, the study of geometry in secondary school stands as a chance to challenge and improve those conceptions of figure. We will want to say that the study of geometry in secondary school will bring them to a different kind of knowing about figures—hopefully caused by stretching or questioning the conceptions of figure they bring with them from earlier schooling. We will explain this position by elaborating on what we mean by *conception* and by identifying basic conceptions of figure that students bring with them to secondary school.

2.2. Conceptions of Figure: What We Mean by *Conception*

We adopt the notion of *conception* proposed by Balacheff and Gaudin (2010) whereby a conception is "the state of dynamical equilibrium of an action/feedback loop between a learner and a milieu under proscriptive constraints of viability" (p. 213). A conception is thus ascribed by an observer to the learner-milieu system on account of the observation of the system's stable interactions. To identify a particular conception, such as the ones we describe below, would therefore mean to name a stable system of interactions between a knowing agent (e.g., a child) and a milieu (the mathematical counterpart to the child in those actions; e.g., a set of diagrams, a set of tools to make diagrams; see Brousseau, 1997, pp. 51–53). To proceed beyond the mere naming, Balacheff and Gaudin (2010) propose a way of modeling conceptions that consists of accounting for the elements of a quadruplet (P, R, L, Σ). In this quadruplet, P is a set of problems in which the conception is operational (problems that may be written or oral statements in the milieu, or goal states ascribed to the agent). The symbol R represents set of operators used to pursue those problems, and can be interpreted as schemes or operational invariants ascribed to the agent or as repeatable, transposable actions that the agent is observed doing with or without instruments. The symbol L denotes the semiotic register or system of representation employed in communicating the problem to the agent as well as used by the agent to act on the milieu and to read the feedback from the milieu; it may be a language, another sign system (by saying "sign system" we allow that it will have meanings as well as ways of combining those), or the semiotic elements of a tool (e.g., the menus and icons for functions in DGS). Finally, Σ is the set of controls inscribed in the milieu or ascribed to the subject on account of observations of their decision making; controls check for the correct application of operators and for whether a given problem has been solved correctly.

Such a way of accounting for conceptions keeps them tied to the spheres of practice in which they manifest, and thus permits us to associate different conceptions to the same knowing agent: Since conceptions are ascribed to the interaction of a subject and a milieu, it is reasonable that, when immersed in different situations (facing different milieux, working on different kinds of problems), the same person might respond in ways that an observer might judge as evidence of different, conflicting conceptions, even if the observer might use the same mathematical ideas to describe both situations. For example, Schoenfeld (1988) showed how students could be observed using different controls for their problem solving when working in situations of construction and of proof. Similarly, González and Herbst (2009) used the notion of conception to illustrate how milieux with and without dynamic geometry software could make room for quite different conceptions of congruence. The notion of conception thus makes it possible to describe a person's apparently contradictory ways of knowing (observed at different times) as different conceptions. The notion of conception can be used to provide practice-centered accounts of mathematical knowing, and in particular it can be used to describe conceptions of figure. More broadly, the notion of conception can account for a host of related phenomena that have been studied in mathematics education, including bugs, errors, intuitions, misconceptions, naive ideas, and obstacles; while in this chapter we use it to account for some curricular features of secondary geometry, the notion of conception is also useful to organize what the field has found about students' thinking and learning and about the potential impact of teaching interventions on students' learning.

The notion of conception can be understood as a fractal object of sorts: It provides the means to model the mathematics embedded in practices at various levels of mathematical zoom. To illustrate this point, and in spite of it being a nongeometric example, consider the relationship between *function* and *linear function*: Mathematically, the latter is a particular case of the former, but the notion of conception can be used to analyze both. On the one hand, Balacheff and Gaudin (2010) describe two conceptions of the mathematical notion of function (in general): The *curve-algebraic* and the *algebraic-graph* conceptions. But the notion of conception could also be used to describe conceptions of linear function or of its component elements such as slope (e.g., Stanton and Moore-Russo, 2012; Stump, 2001). Bearing in mind this fractal image of conceptions, we develop the notion of conceptions of figure and identify particular conceptions of figure, keeping open the possibility that in the study of phenomena that might be modeled with particular figures (e.g., the rectangle) one might still uncover different conceptions of that figure.

As noted above, as regards a particular notion of reference (e.g., function, rectangle), the same cognitive agent (e.g., a student, a group of individuals working together) might enact different conceptions at different moments in time. The set of all those conceptions in which the agent participates within a period of time is referred to as the agent's *knowing* of that notion of reference (noting that such a set might contain inconsistencies), while the set of all such knowings (e.g., by different individuals, or by different groups) is referred to as the *concept* of that notion of reference (Balacheff and Gaudin, 2010), which is compatible with Tall and Vinner's (1981) notion of the "concept image" (discussed in Chapter 3).

It is worth stressing that conception, knowing, and concept, denoted by Balacheff and Gaudin (2010) as ck¢, are epistemological rather than psychological constructs. That is, they describe what is known by the agent, but not how the agent comes to know it, nor how the agent mobilizes his or her organism to act in a way that reveals such conceptions or knowings. Yet these epistemological constructs have enough diversity to locate what is known not in any abstract realm of disciplinary knowledge but in the practices that humans engage in. The notion of conception relies on scholarship from various branches of psychology; the notion of operational invariant or scheme, coming from Piaget's work used to describe a conception, is an example. While this epistemology uses classes of problems as anchors for conceptions, the reader should not assume that such problems are bearers of an intention to teach or that they necessarily enable the development of conceptions—they are merely sites where the conceptions might be at work.

In what follows we build on the epistemological perspective brought by the ck¢ model of conceptions to identify four basic conceptions of *figure* that we expect describe students' earlier geometry experiences. As we move to conceptualizing the study of geometric figures in secondary school, these basic conceptions of figure provide the background against which secondary geometry experiences can be organized. The reader could expect that each conception of figure might be modeled using the (P, R, L, Σ) quadruplet. We will suggest starting points to such modelization, leaving the details for the reader.

2.3. Initial Conceptions of Geometric Figures

This section introduces conceptions of geometric figure that could be associated with students' early experiences with space and shape in and out of school. Students deal with space and shape from very early in life: Playing with toys, climbing playground structures, and moving about the house or the city, with some of these experiences being mediated (e.g., riding cars, using levers, using language). We describe four conceptions of geometric figure, four sets of practices that, we argue, we can typically count on children having participated in by the time they come to secondary school. These are very general conceptions of figure, though we illustrate them using particular examples. In keeping with the notion that conceptions are entities of a fractal nature, each of these conceptions of figure can likely be broken down into conceptions of specific figures.

A first approximation to identifying these conceptions is the realization that children have experiences in three kinds of spaces, distinguished among them by their scale in relation to the human body. Berthelot and Salin (1998, p. 72) note

> Because of the constraints due to the size of the different spaces in which daily life interactions take place, natural knowledge of space is strongly structured into three main representations: microspace (corresponding to the usual prehension relations), mesospace (corresponding to the usual domestic spatial interactions), and macrospace (corresponding to unknown city, maritime, or rural spaces).

Laborde (2000, p. 133) also alludes to the importance of distinguishing these three spaces in regard to conceptions when she indicates that "Evidence has been gathered showing that children principally develop controls and spatial knowledge of a small-size space (microspace, the space of objects that they can take in their hands and manipulate)" but that students "have inadequate representations of both mesospace (from 0.5 to 50 times the size of an individual) and macrospace (in which objects cannot be grasped by hand or eye, e.g., the space of the city)".

These three kinds of space are of interest for two reasons. On the one hand research (Acredolo, 1981; Berthelot and Salin, 1998) has shown not only similarities but also differences in spatial performance when working on a problem in spaces of different scale. Thus the distinction seems quite opportune when searching for different conceptions. On the other hand, the emergence of new technologies such as handhelds and Geographic Information Systems (GIS), and the democratization of access to other technologies such as photographic cameras, architectural design software, or drones has brought important resources for us to think anew about the role that experience with space and shape can have in the abstraction of geometric models of the experienced world. This in turn is, as we anticipated, the direction we would like to propose for thinking about the study of geometric figures.

We identify four conceptions of figure related to those three spaces of activity. We make no claim of completeness; we merely propose these four conceptions as starting points that might articulate the geometric knowings available to school students as they move into secondary school. It should be clear to the reader that we are organizing elementary conceptions from a perspective that is serviceable to the teaching and learning of geometry to adolescents; scholars who focus on the geometric thinking of children might find other ways of organizing children's conceptions more compelling (e.g., see Battista, 2007).

2.3.1. The Figure as Navigation of the Macrospace

By the *macrospace* Berthelot and Salin (1998) allude to the space of regions, cities, and building and ship interiors; crucially, none of those spaces can be captured in their totality by gaze. This is what Acredolo (1981) had called the large scale space. We take that to mean the larger spaces that contain us, including buildings, streets, parks, hiking or biking trails, the sea, the countryside, etc. From an observer's perspective, we know that this physical world can be described using geometry: The notion of angle can help describe the relative position of two streets or the steepness of a hike; the notion of path length on a sphere can help make sense of alternative routes from one place to another. Young students live in and experience that macrospace, and as they do so they utilize abilities of *spatial orientation*, which Clements (2004) defines as "knowing where you are and how to get around in the world, that is, understanding and operating on relationships between different positions in space, especially with respect to your own position" (p. 278). Children navigate the macrospace and in doing so they rely on conceptions of figure that help them deal with location, distance, amplitude, turn, and so on. In general, to the extent that children can manage their movement in the macrospace, one could hypothesize that they act within a conception of figure as (an accomplished or planned) navigation of the macrospace (see Kotsopoulos et al., 2015; Lee et al., 2012). This conception would likely expand into several conceptions; for example, it could be expanded to distinguish between the navigational geometry of sighted and sightless children (Landau, 1988). Literature in psychology has dealt at length with spatial orientation in early childhood and cannot be summarized here; the reader can consult Liben (2007), Newcombe and Huttenlocher (2000), and Piaget and Inhelder (1963) for in-depth discussions on these matters.

Following the lines of our earlier fractal analogy, we may unpack the conception of figure as navigation of the macrospace into conceptions of specific navigation figures. Our goal here is only to illustrate what kinds of figures we are talking about; the literature sources noted above, in particular Clements and

Sarama (2014) can be consulted for more specific details. To exemplify how the conception could be characterized using the quadruplet (P, R, L, Σ) we invite the reader to consider the problem of ascertaining one's location in the macrospace (for example, in order to report it to others). Newcombe and Huttenlocher (2000, pp. 13–22) identify four types of spatial coding that could be considered clusters of operators used for location problem solving: Cue learning (association with landmarks), place learning (orientation and distance from landmarks), response learning (explicit memory of motor movements), and dead reckoning (a.k.a., inertial navigation or path integration, which uses tacit memories embodied in kinesthetic and proprioceptive mechanisms of the agent; see also Sarama and Clements, 2009, pp. 163–166). Clearly those operator clusters can be further analyzed into behaviors and sensations, but the reader may already get the idea that these operators include bodily functions such as walk, increase or decrease step size, look, listen, jump, turn a given amount and direction, etc. Operators might also include counting, marking, and measuring time, depending on which type of spatial coding is used. In looking for the semiotic register we ought to ask what communication modalities young children use (Streeck et al., 2011). For this conception of figure, the semiotic register includes the various modalities that those bodily functions utilize, such as written and oral language, gestures (e.g., pointing), as well as conventional signage such as street and traffic signs. Controls can refer to (1) controls of appropriate application of an operator, such as 'if a still object is to my right when I walk north, it should be to my left when I walk south;' (2) controls for the reading of messages such as 'if someone says to 'turn right' and they are not oriented the way I am, I'd better ask 'my right or your right''; and (3) controls for the solution of the problem, such as when going to a desired location through a chosen path, realizing that one is closer to the target place because objects have increased in apparent size.

2.3.2. The Figure as Capture of Objects and Actions in the Mesospace

The mesospace consists of the space of objects that can be captured by gaze, objects whose size, according to Laborde (2000), ranges from about half the size of the agent to 50 times its size. We include here objects like cars, refrigerators, tables, benches, exercising equipment, and construction materials, and the actions that individuals do with and around these objects. Young students can observe and manipulate those objects but the evidence seems to indicate that they do not do so in the same manner as they do with smaller objects: Berthelot and Salin (1998) report that students who knew about properties of rectangles, failed to control for angles when asked to map the location to which a rectangular bench would be moved. A conception of figure as observable and manipulable rigid object could be hypothesized as one that facilitates children's interaction with the mesospace.

In this conception of figure as capture of large objects, a problem might include the need to describe a visible object (such as the bench in the experience described by Berthelot and Salin, 1998) in such a way as to anticipate the space it will take when moved to a new space. Operators might include measuring dimensions of the object—for example, measuring linear dimensions with a tape measure. The semiotic registers include written and oral language (where names of geometric objects are used to describe objects or their parts), a pictorial register (where versions of geometric diagrams are included), and the numbers and units used to communicate dimensions. Controls can refer to (1) controls of appropriate execution of an operator, such as how to make sure the measuring tape is positioned correctly; (2) controls of the reading of a message, such as what to look for when reading from the positioned measuring tape; and (3) controls that the problem has been solved, such as what measurements are needed in order to be able to anticipate the location of the object. As in the case of the conception of figure as navigation of the macrospace, in this conception of the figure as visualization (grasping) of objects in the mesospace, zooming in further into particular conceptions is more likely to reveal substantial differences between conceptions of figure at different scales. In regard to the conception of angle, for example, Berthelot and Salin (1998), Devichi and Munier (2013), and Fyhn (2008) suggest different representations of angles in the mesospace, which are not always amenable to treatment in the ways of relating to angles at the much smaller scale of the microspace.

2.3.3. Conceptions of Figure in the Microspace

In primary school, children have more experiences with the microspace than with either the macrospace or the mesospace. There is also much more research on their understanding of geometric concepts as realized in the microspace (e.g., Battista, 2007). From the conceptions perspective, we distinguish conceptions of figure in terms of the practices in which the agent interacts with these objects. Along those lines we may distinguish between two conceptions of figure in the microspace: Figure as construction of small objects, and figure as description of small objects. We include in the former experiences distorting small figures, and in the latter experiences displacing small figures. This distinction between these different conceptions of figure is only provisional, and we suspect that as one zooms into conceptions of specific figures, description and construction might have nonempty intersection. But because early on they place radically different demands on occupational (particularly manual) dexterity, we separate these into two basic conceptions.

2.3.3.1. The Figure as Construction of Objects in the Microspace

The first conception of figure in the microspace is that of *figure as construction of small objects*. This conception refers in general to the practices that involve creating, drawing, or altering by breaking or distorting and recomposing small objects, including in particular diagrams and replicas,[1] as in when a child uses geometric shapes to create a composite figure that is given to them in pictorial form, and especially when a child draws diagrams or makes models with loosely structured materials (e.g., pencil and paper, toothpicks and marshmallows). Early on children fill simple empty frames by using shapes individually, but later they become able to join shapes together to compose what the empty frame calls for. Working with tactile materials and progressing toward filling more complex frames, early modes of operating by trial and error later progress to an ability to anticipate how a side length or angle might fit if it were to be rotated or reflected. Making determinations as to whether a shape fits the frame indicates the child's control structure in which the lines that make up the frame give feedback to determine whether the shape fits. While, early on, children receive feedback from the figure only through visual discrimination, later the child can construct a mental image of what might occur with a given reflection or rotation (Sarama & Clements, 2002, p. 98; see also Bruce & Hawes, 2015; Clements et al., 1996; Mansfield & Scott, 1990; Ng and Sinclair, 2015). This process may include attention to the properties of a shape; however, they may only focus on one property of the shape at a time, such as side length or trying to fit corners (Sarama & Clements, 2002, p. 98). Through much experience making use of such schemes, children become more intentional in regard to this task, in which rotations or flips are used as a way of choosing shapes that attend to several constraints (e.g., side length and angle) simultaneously. This gives way to a student's ability to form composite shapes (Clements et al., 1997) that are used as substitutions for other shapes or as an iterative tool for filling or patterning.

Thus in this conception, problems may include the creation of a diagram or a new physical model of a figure given by its description (e.g., by its name or properties) or by a representation (e.g., a picture or replica). Operators include a number of occupational tasks done with small objects, such as creating strokes with pencil on paper, folding, holding steady, cutting, measuring length with ruler, measuring angles with protractor, creating arcs with a compass, glueing, moving the mouse on a mousepad, etc. Semiotic registers include here the pictorial, spoken, tactile, kinesthetic, or written signs that may be used to provide a figure to be constructed and to read information about the figure to be constructed, as well as retinal images of the constructed figure, including those on a computer screen. Controls may include (1) controls of correct application of an operator, such as 'the stroke through two dots is a line if it superimposes with the rim of a straightedge', (2) controls of the reading of a message, such as 'the shape representing the side of a cube seen in perspective appears like a parallelogram even though it is a square', and (3) controls that the problem has been solved, such as "if the constructed figure looks like the given model, then the construction has been completed."

2.3.3.2. The Figure as Description of Objects in the Microspace

The second conception is that of *figure as description of small objects*. We include here the practices of naming and describing using language, as well as the practices of measuring and describing using numerical attributes and the practices of describing the displacements of a figure (see Bautista and Roth, 2012; Sack, 2013). Problems in this conception include matching a shape or object with a name, stating properties of a given shape or object, recognizing a shape or object given its properties, relating two shapes or objects, and moving a shape or object to a new location. Operators include looking, measuring lengths, measuring angles, calculating with numbers, reading, ruling out, etc. Semiotic registers include the written and spoken language, the pictorial register (particularly the diagrammatic register described by Dimmel and Herbst, 2015a). Controls include (1) controls of correct application of an operator and (2) controls that the problem has been solved.

Clements and his colleagues (1999) have engaged students in problems that call for this kind of activity. In this context, when children are asked to explain their reasons for identifying a particular shape, their initial descriptions are based less on the properties of geometric figures and more on their experiences with shape in the world. They might say "it looks like this" to compare a figure to something they have in their visual field. In doing so they tend to describe the figure based on a type to which they have compared, but without attention to particular features of the shape. Or, they may indicate some aspect of the shape that shows they are attending to an earlier sense of properties, as when a child says a shape is 'pointy' to indicate attention to a vertex (Clements et al., 1999). While the former description is based on comparison to a type, the latter is an indication of early attention to geometric properties in which explanations are based on informal language or understanding of the properties themselves. Not until they are older can children attend to properties separately as they consider how each property plays a role in the figure. Whereas younger children do not focus on single features, but rather make comparisons to a type of a similar figure, older children attend to specific features that are known to make figures of that type (Clements et al., 1999).

2.3.4. Conceptions of Figure and the Secondary Geometry Curriculum

Our goal in this section has been so far to use the notion of *conception* as sphere of practice to identify four spheres of practice in which children have dealt with geometric figures. Clearly these four conceptions have merely been identified, not completely characterized; this would be beyond the scope of this book and the interested reader can consult, for example, Clements and Sarama (2014) and Sarama and Clements (2009) to learn more about specific studies in early childhood geometry that could help develop such characterizations.

We claim that when students arrive in secondary school, they already know geometric figures in ways that can be described in reference to those four different conceptions: As enactments in the macrospace, as captures of objects in the mesospace, as descriptions of objects in the microspace, and as constructions of objects in the microspace. These conceptions ground the experiences that students could have with figures in secondary school.

At the secondary school level students have the opportunity to make those ways of knowing geometric figures more complex or to use those prior conceptions to build more sophisticated conceptions of figure. We bring those conceptions in here not so much to report on what they are, which would likely require a book of its own, but as points of departure for a conceptualization of what the secondary geometry experience could be. In this chapter, this conceptualization takes the shape of a consideration of the geometry curriculum and the conceptions approach plays a fundamental role.

2.3.4.1. Modeling the Experiential World

A consideration of the curriculum in terms of conceptions, that is in terms of the spheres of practice in which students are involved, helps bring in, for the case of their practices with space and shape, the notion that knowledge of geometry could emerge as students progressively *model* their *experiential world*. We mean *experiential world* to include the practices in which students are involved: Real world experiences with space and shape are indeed included but they are not the only ones; we also mean experiences with representations (e.g., experiences reading or drawing diagrams) and, in general, we mean experiences in the various practices where they have developed conceptions of figure. We mean the word *model* in a general mathematical sense of a representation with which it is possible to do some kind of calculation (logical, numerical, algebraic) in order to anticipate (or predict) experience. The discipline of geometry presented through modern axiomatics such as Hilbert's offers a possible horizon for that modeling approach (see Figure 0.3); Hilbert accounts for one possible mathematical model of the human experience with figures. But rather than thinking of secondary geometry instruction as the delivery of that or any other accomplished model, we think of the secondary geometry instruction as the creation of conditions that permit the development of models that might eventually aspire to become that one. We take the symbolic representation and axiomatic-deductive means of calculation from that model as suggesting the horizon to which those models could aspire.

In particular, we recall the words of Freudenthal, who regarded geometry as "one of the best opportunities which exist to learn how to mathematize reality" (Freudenthal, 1973, p. 407), and we argue for a secondary geometry curriculum that provides the means for students to theorize their prior experiences with figures. In particular, we argue for the value of investigating a secondary geometry curriculum that provides students with, on the one hand, experiences in which they have to predict whether actions in the experiential world will or will not work and to explain why those actions may not have worked and, on the other hand, the means to handle such kind of questions. Existing conceptions of figure provide accounts of the experiential world within which such modeling activity takes place. The semiotic registers and controls within each of those conceptions provide the building blocks for the initial models that students might use in such experiential worlds.

The Euclidean diagram is a canonical example of one such model. Euclidean diagrams make use of a semiotic register (i.e., signs that function as "lexical" elements, combined according to a "grammar" of sorts that allows them to make available meanings) as well as controls (i.e., ways of validating or invalidating those available meanings). To substantiate this point we look into the existing research on the role of diagrams in secondary school teaching and learning of geometry and on neighboring areas of cognitive science, history, and philosophy. We take diagrams as one important element of the system of representation in all those conceptions of figure and we use it as a privileged case of the larger class of concrete representations of figure on which a transition to theoretical work in geometry could be anchored. In the following we review the most important points from that research that contribute to ground our way of thinking about the geometry curriculum. We then consider how it is possible to use those existing conceptions to develop in students a theoretical stance toward geometric figures, and eventually a knowledge of geometry as a mathematical theory.

2.4. The Geometric Diagram in the Literature

In this section we weave contributions from the literature on diagrams to suggest that the work of moving from the four basic conceptions to more sophisticated conceptions of figure can be anchored in activities that use and further problematize the representations of those basic conceptions of figure. Among those representations, a notable one on which there is substantive scholarship is the geometric diagram, but we take this review as possibly relevant to other comparable representations such as maps, replicas, photographs, and dynamic sketches. We lead this review with the question: What are the affordances and constraints of diagrams in the study of geometric figures?

2.4.1. The Geometric Diagram in Mathematics Education and Cognitive Science

Many mathematics education scholars have contributed to scholarship on diagrams and their relation to geometric figures, sometimes referring to them as "pictures" or "sensory representations" (e.g., Duval, 1995; Fischbein, 1993). For more comprehensive reviews of this literature we encourage the reader to see Battista (2007), Clements and Battista (1992), and Sinclair et al. (in press). Here we concentrate on some that have been particularly helpful to make our point.

2.4.1.1. Seeing vs Knowing: How Diagrams Preserve and Lose Information

In his studies of students' creation and interpretation of representations of three-dimensional figures, Parzysz (1988) argued for the need to distinguish qualitatively the two dimensional diagrams of plane figures from the two dimensional diagrams of three-dimensional figures. Parzysz (1988) reserved the word *figure* for the mathematical object of reference, which could be either three-dimensional or two-dimensional. In thinking about the representations of those figures he distinguished between two kinds of representations for the case of three dimensional figure: the *model* and the *drawing*, which he judged different in regard to their informational content. Parzysz (1988) proposed that the model of a three-dimensional figure, a three-dimensional physical representation, what we call a replica, stands in relation to the three-dimensional figure as the drawing of a two dimensional figure stands in relation to the two-dimensional figure, and he showed that a drawing of a three dimensional figure necessarily loses information to the point of inducing a dilemma in students between *knowing* and *seeing* the properties of three-dimensional figures (see more on this in Chapter 3). Parzysz (1988) related this dilemma between knowing and seeing to a contrast between different movements in art history: On the one hand, the classical development of perspective drawing, which enabled the creation of representations in which the properties of three-dimensional figures could be *seen*, and on the other hand, the Egyptian "aspective" (Johnson, 2002; see also Brunner-Traut, 1974/1919) drawing, which would require the observer to cope with the limitations of the drawing by *knowing* what the properties of the figure should be (because they are not visually preserved). Parzysz's (1988) distinction between *model* and *drawing* starts to create a bit of texture within the class of things we have called *diagrams*, though we are not ready to adopt the words *model* and *drawing* in the way Parzysz (1988) suggested. Regardless of word usage, a central contribution from Parzysz (1988) is, however, the notion that different representations of a figure can be contrasted to one another in regard to the ways in which they visually preserve or lose information. Further, the distinctions in visual geometric properties between art movements (i.e., aspective vs. perspective art[2]) highlights that this preservation or loss is not intrinsic to a representational medium (2D vs 3D), but only perhaps constrained by the medium and dependent also on general principles used in making the representation.

2.4.1.2. Spatio-Graphical vs Theoretical Properties: How Diagrams Add Information

Laborde (2005) made a contribution that complements Parzysz's (1988), when she proposed the need to speak about two kinds of properties of geometric figures, by distinguishing the types of information conveyed by a diagram. To substantiate the more general point that seeing a diagram can be an obstacle for knowing a figure, Laborde (2005) noted that a diagram such as that of a parallelogram offers an array of concomitant *spatio-graphical* properties. Some of those spatio-graphical properties, such as the color and width of strokes, are not geometrical (though they can be signs that point to geometric meanings; see Dimmel and Herbst, 2015a). Other spatio-graphical properties, such as the specific length of segments or the specific amplitude of angles, are geometrical but incidental to a specific parallelogram—a parallelogram could have such side lengths and angle sizes but could also have others. Finally, other properties are what Laborde (2005) calls *theoretical*:

They are properties necessitated by the definition of parallelogram (e.g., opposite sides have the same length). This distinction between spatio-graphical and theoretical properties thus suggests that (pace Parzysz, 1988, who spoke of preservation and loss of information) diagrams also *add* information: They may not only represent what is necessarily true about the figure but also truths that are incidental to the figure. This is the case not only with two-dimensional diagrams, be those sketched or constructed with pencil on paper or with DGS on the screen, but also with physical objects that represent three-dimensional figures. Both Parzysz (1988) and Laborde (2005), each in their own way, point to the opacity of geometric diagrams as representations of geometric figures: They may fail to represent what is necessarily true and they may also represent what is not necessarily true.

2.4.1.3. The Epistemological Need for Representations

It might be tempting to suggest that geometry should be taught without diagrams. The ferment around axiomatics that emerged toward the end of the nineteenth century through the work of Peano and Pasch and that culminated with Hilbert's *Foundations of Geometry* (see Corry, 1996, p. 155; Royce, 1905), prompted some mathematics educators such as Halsted (1893) to make such a suggestion; it reached a pinnacle with Bourbaki (see Sinclair, 2008, p. 49). Yet, Fischbein's (1993) notion of *figural concept* complicates matters with respect to understanding how the diagrammatic representations of geometric concepts afford them possibilities that are not entirely contained in their abstract definition—possibilities that are particularly useful for understanding the necessity of theoretical properties. For example, Fischbein (1993) illustrates how some metaphorical operations on diagrams (e.g., rolling a circle, flipping a triangle, superimposing figures) are on the one hand conceivable because the concepts can be diagrammed, and on the other hand useful to prove that some properties of those concepts are necessary. While some spatio-graphical properties of diagrams are not theoretical properties of concepts, the fact that those concepts can be diagrammed (and that for that purpose they end up having some spatio-graphical properties) enables the deduction of theoretical properties.

2.4.1.4. Diagrams as Representations of Many Concepts

Larkin and Simon (1987) suggested a further complication in their comparison between sentential and diagrammatic representations of information and their role in problem solving. Larkin and Simon (1987) studied the processing involved with both geometric diagrams and physics diagrams (pulley systems) and their sentential counterparts. Their main finding was that both kinds of diagrammatic representations required less processing than their sentential counterparts. But in describing the geometry comparison as more complicated than the physics comparison, Larkin and Simon (1987) noted an important extra feature of sententially-stated geometric problems that is relevant when reading geometric diagrams: In sententially-stated geometric problems "the given data structure does not match the given program. Therefore to solve the problem at all, the problem solver must enhance the data structure in ways that prove considerably easier with diagrams than with sentences" (Larkin and Simon, 1987, p. 82). By the "given data structure" Larkin and Simon (1987) meant the givens of a problem (e.g., "Two transversals intersect two parallel lines and intersect with each other at a point x between the two parallel lines", p. 82), and by the "given program" they allude to the statements of definitions and properties that can be used in solving (e.g., "ASA. If two angles and the included side of one triangle are congruent to the corresponding two angles and included side of another triangle, then the triangles are congruent," p. 83). Larkin and Simon (1987) explain the extra work needed by the sentential solver of the geometry problem after identifying the givens: "But then we are asked to prove congruent two triangles that have not been mentioned. This situation already contrasts with that of the pulley problem, where every necessary element was specified in the problem statement. In just interpreting the problem statement we already have significant difficulty in recognizing all mentioned elements" (p. 83). The geometric diagram dramatically reduces the extra work of recognition that the solver would need to do when working out a sententially-stated geometry problem: The diagram not only shows together things that are related (as the pulley diagram in the physics problem does), but it also makes apparent the different kind of objects that could be read in the data (e.g., three intersecting lines are not only that but they also make a triangle). Thus a geometric diagram facilitates choosing how to encode what is being given.

Coliva's (2012) examination of "seeing as" provided a conceptualization of the phenomenon noted by Larkin and Simon (1987). Elaborating on Wittgenstein's (1958) notion of *seeing as*, Coliva (2012) has provided a disciplined philosophical analysis of *seeing-as* in geometric diagrams, contrasting it with related concepts such as *inferring*, *seeing*, and *taking-as*. As Coliva (2012) notes, geometric diagrams often offer the immediate opportunity to see geometric objects *as* one object (e.g., a configuration of three pairwise intersecting lines) or another (viz., a triangle). In contrast, a sentential representation would require an elaboration of the given data to infer from the existence of three intersecting lines the existence of a triangle. Coliva (2012) contends that this seeing-as affordance of geometric diagrams, in which objects can simultaneously be seen as instances of many concepts, is key in explaining the role that diagrams might play in Euclidean geometry proofs.

In a related vein, Duval (1995) noted that geometric diagrams (which Duval calls *geometrical figures*) may be apprehended or grasped cognitively in four different ways. A *perceptual* apprehension is needed in order to distinguish the figure from its background and the parts from the whole, recognizing figures and subfigures pictorially and perhaps being able to describe them using their conventional geometric names. A *sequential* apprehension is needed in order to construct or reproduce a figure using tools. Construction and reproduction require the cognizing agent to decide what is drawn first and what later (perhaps on the basis of what was constructed before), and in that sense it calls for a temporality that is not needed in the perceptual apprehension. Duval (1995) calls the third kind a *discursive* apprehension, referring to how discourse (in the sense of either oral or written discourse) that names a set of properties also chains them in a structure of dependencies in which, for example, some are taken as hypotheses and others as conclusions (as when one incorporates them in an if-then clause). Finally, the fourth kind is an *operative* apprehension, by which Duval (1995) alludes to a number of possible ways of seeing a diagram as a class of diagrams, either by noticing different decompositions of it, or by noticing different locations or positions of it, or by noticing different sizes of it. Duval's (1995) four kinds of apprehensions move us beyond the diagram and to the realm of thinking about the diagram, which is addressed in more detail in the next chapter. Yet they are included here to suggest that the seeing-as noted by Coliva (2012) is only one of a number of ways in which a diagram is amenable to making multiple meanings.

As one considers the insights from Duval (1995), Larkin and Simon (1987), and Coliva (2012) along with the contributions from Parzysz (1988), Fischbein (1993), and Laborde (2005), the panorama might become even more complicated. Following Parzysz and depending on what is involved in the representation (e.g., techniques of perspective drawing, tools), the diagram might preserve, degrade, perhaps even lose some of its affordances, including how well it supports seeing-as (which is a case of Duval's operative apprehension) as well as other apprehensions. One might also ask whether these apprehensions further complicate the phenomenon noted by Laborde: Does seeing-as validate spatio-graphical properties that might not be theoretically provable? We come back to cognitive issues related to diagrams in Chapter 3 but first draw on philosophical and historical scholarship to inspect how the use of diagrams has been key in the validation of geometric properties that are true. To the extent that we can know about historical practices of diagramming, we would also want to know whether those historical practices can inform the way we engage students in diagramming in classrooms.

2.4.1.5. Many Diagrams as Representations of Single Concepts

All the discussion above has taken as given the notion that the concept of a figure is instantiated in an object represented by a single diagram—where the distinction between theoretical and spatiographical properties can be mapped onto the distinction between generic objects and particular objects. In that world of single or few geometric objects, the formulation of properties of concepts might be mediated by taxonomies and case studies (as when properties are studied separately for acute, right, and obtuse triangles, taking single diagrams that represent each case); within each case, generalization of properties might be done by deduction or abduction. But in recent years DGS has enabled the consideration of digital diagrams where one can (1) drag points on the screen to change one diagram into another; (2) animate points in a diagram so as to represent a class of diagrams on the screen; (3) construct diagrams so as to have specified properties which are then maintained as points are dragged or animated; and (4) continuously measure attributes of parts of diagrams as those parts are dragged or animated.

These dynamic diagrams have enabled radically different ways of relating concepts to objects, including the possibility of conjecturing general properties by recurrence: a multiplicity of different diagrams, associated to each other by dragging or animation, and the set of measures of their attributes associated to each other by inspection or calculation, can lead to the formulation of empirical conjectures about geometric concepts. González and Herbst (2009) called this interaction with diagrams "functional". While all that has been discussed above about static diagrams still applies for each object in a collection of dynamic diagrams, the collection of dynamic diagrams itself affords an empirical distinction between theoretical and spatio-graphical properties of a figure: Theoretical properties are manifest in invariants observed among their recurrent measures, recurrent calculations among measures, or recurrent spatio-graphical properties (see also Hölzl, 1996; Laborde, 2005; Laborde, et al., 2006).

2.4.2. The Crucial Role of Diagrams in the History of Geometry

In spite of the complexities associated to diagrams, diagrams have been crucial in the history of geometry. "Demonstrations in Euclid's *Elements*, from Proposition 1.1 on, use their diagram essentially to introduce items such as that notorious intersection point of the two circles, for which Euclidean demonstration has no alternative justificational resources" (Manders, 2008, p. 65). Indeed, while Euclid is (as discussed in Chapter 1) usually praised for having written the first systematic exposition of mathematics in which postulates and definitions were explicitly laid out and then propositions were stated and proved on the basis of those postulates and definitions, it is also well known that Euclid's postulates were far from complete (Hartshorne, 2000b). In particular, Euclid did not include postulates for continuity, order, and separation, even though he took for granted tacit properties of that sort, as Manders (2008) notes in the quote above, relying on the diagram to communicate their warrant. For example, Euclid's Proposition 1.1, which shows how to construct an equilateral triangle given its side, relies on the diagram to show that two congruent circles, each of which has points inside and outside the other circle, will intersect. Euclid's reliance on diagrams to warrant tacit properties of continuity, order, and separation did not prevent him from elaborating much of the knowledge of geometry (Panza, 2012), but those omissions were later, especially in the nineteenth century, noted as shortcomings. In the second half of the nineteenth century, through the work of Pasch and especially of Hilbert, those missing postulates were made explicit and Euclidean geometry was established on an axiomatic basis so that its propositions could be proved without recourse to diagrammatic reasoning (Greenberg, 2010). Hilbert's *Foundations of Geometry* has been highlighted as a way of vindicating Euclid by adding what he missed (Greenberg, 2010; Mueller, 1981).

A notable accomplishment of Euclid's *Elements*, in spite of his reliance on diagrams and lack of axioms, is that the conclusions he arrived at are true. Manders (2008) notes that "ancient geometers achieved their lasting, subtle, and powerful results precisely by the means that philosophers dismiss so high-handedly today, without the benefit of modern logic and Hilbert's refined control of coordinated domains" (p. 67). To explain the efficacy of Euclidean geometry in spite of its modern shortcomings, Manders (2008) contributes two important points.

The first of Manders' (2008) points is that the canonical reading of Euclidean diagrams is one that happens along with the reading of the text of the proofs, as if one were following sentences in the proof with corresponding acts of noticing in the diagram. The argument for the canonical reading of the diagram finds support in Reviel Netz's (1998a) study of the lettering of classical Greek diagrams. Netz showed that, in addition to points that were *determined* by the text (e.g., "let *A* be the center of a circle"), diagrams would also contain two other kinds of labeled points. Diagrams would contain labeled points that were *underdetermined* (i.e., one could not completely determine them from their textual introduction) and labeled points that were *undetermined* (i.e., they had no textual introduction). In regard the latter, the text of the proof would speak about them using a letter but to understand what the text meant by the letter one would have to recur to the diagram and see where the letter was. Furthermore, Netz (1998a) suggests that the labeling sequence of such points is evidence that the writing of the proof was preceded by a sort of dress rehearsal, an oral argument made between the prover and the diagram, where, we speculate, other elements (e.g., auxiliary constructions) would be drawn and points needed in the argument would be labeled. All this takes Netz (1998a) to suggest that letters, in Greek diagrams and in the text of proofs, were not symbols of mathematical objects but rather indices of objects: Letters pointed to objects which the diagram (not the letter) represented. Thus the diagram was not merely an illustration of what was being represented by symbols in the text of the proof; rather the diagram was part of the representation itself, partaking with the text in the construction of a proof for a geometric figure. Note then, following Laborde (2005), that the representation system used by Euclid could potentially be fraught with the capacity to communicate unintended messages—it afforded the possibility to read spatio-graphical properties. Or did it?

The second of Manders' (2008) points helps understand how proofs about a particular diagram might have succeeded in establishing results that were general. Manders (2008) proposes a distinction between *exact* and *co-exact* properties. Co-exact properties are those that would not be falsified by small alterations of a diagram. Manders illustrates this with the circles in Euclid's Proposition 1.1; the circles centered on extremes of a segment intersect, regardless of how long we draw the segment. Exact properties, however, are those that could get falsified by alterations of the diagram: Claims of congruence and parallelism are examples of exact properties. Then Manders (2008) claims that the text's dependence on messages communicated by diagrams in Euclidean geometry is restricted to co-exact properties, which seems to explain the capacity for Euclidean geometry to establish general results. The co-exact properties of diagrams may not have been symbolized and axiomatized to be treated in the geometric text with the tools of deductive logic, but they were represented in diagrams (and those diagrams pointed to with letters used as indices) so that they could be talked about and reasoned with. As Netz (1998a) says, mathematics is "the field in which we apply reason to inter-subjectively definitely given objects (Notice that I insist upon the word 'reason'—I don't think we necessarily apply *logic* in mathematics)" (p. 38, italics in original). It seems that the mathematical work to which the diagrams were put in Euclid's *Elements* was not one that encouraged reading off incidental, spatio-graphical properties, but only those spatio-graphical properties that, eventually, would turn out to be theoretical. The theory for those was not available yet, and would not become available until Hilbert's *Foundations of Geometry*.

But while the axiomatization of the co-exact properties by Hilbert did succeed in turning geometry into a formal subject, where letters could be treated as symbols and reasoning reduced to logic, the motivation for such work was not philosophical or logical, but mathematical: It was brought to mathematicians' awareness when there was mathematical need for them. Nagel (1939) narrates how the seventeenth century's attempts to overcome the limitations of three-dimensional diagrams and to prove properties about the relative position of objects in three-dimensional space led mathematicians to develop a different relationship to geometric notions such as point and line: Starting with Desargues's notion that parallel lines were lines that intersect at infinity, and reaching a pinnacle with Poncelet's use of the principle of continuity to accept the notion of imaginary points, mathematicians were led to a different relationship with geometric diagrams, one that would eventually call for the formalization of co-exact properties.

2.4.3. The Role of Diagrams and Logic in Geometry Instruction

The publication of Hilbert's (1902) *Foundations of Geometry* was the culmination of efforts to formalize the geometry that Euclid had started. Such efforts had started earlier in the nineteenth century (especially through the work of Pasch and Peano; see Kennedy, 1972) and had prompted educators to recommend changes in the curriculum for secondary geometry, attempting to make it more rigorous and less dependent on the diagram to convey information (Halsted, 1893). González and Herbst (2006) have noted that one of the arguments historically used in the US to support the course of studies in geometry has been a formal argument, according to which geometry develops students' competencies in logic (this was also argued in other countries; see Godfrey, 1910). In Halsted's (1893) argument for representing geometric objects with symbols and reducing geometric proof to logic, positing a break between elementary, intuitive geometry and secondary, demonstrative geometry would be a reasonable approach. But as we noted above, in its historical development the discipline did away with the diagram as representation only when mathematical inquiry so required it. There was continuity between the use of diagrams in Euclidean and projective geometry, a continuity that helped ground metaphors such as "intersect at infinity" or "imaginary point," all the while when analytic methods were being calibrated to be able to handle those metaphors with some precision. We argue below that the same can be expected of concrete representations of geometric objects if we take the perspective that secondary geometry consists of the progressive sophistication of students' intellectual means to model, predict, and control geometric diagrams and other geometric representations. Fortunately we are not alone in this proposal.

In the United States, John Dewey (1903) advocated not for a break between intuitive and demonstrative geometry but for a continuous modulation between one and the other because "the high school pupil['s] . . . greatest need is neither merely intuitive nor strictly demonstrative geometry, but rather skill in moving back and forth from the concrete situations of experience to the abstracts in geometric statement" (p. 398). Concurrently in the United Kingdom, Charles Godfrey also wrote that

> all experiment at first, all theory later. This I am sure is a mistaken view, it springs from a deep-seated if undefined belief that mathematical thought is solely deductive, and that any other element in mathematical teaching is a rather disreputable intrusion, inevitable perhaps, but a thing to be got over and done with as early as may be, like measles and mumps. What is really needed is a sensible blend at each stage, and anyone will come to this conclusion who will take the trouble to examine the workings of his own mind when a new study is undertaken.
>
> *(Godfrey and Siddons, 1931, p. 21)*

For Dewey, the appropriate curricular questions about managing such transition were "What factors not valid in a strictly demonstrative sense are needed in order to assist the mind, which is only in the process of becoming logical in general and of becoming geometrically logical in particular, in securing its introduction into, and vital familiarity with, geometrical subject matter?" and "How shall the extra-logical factor be so introduced as not to leave mental habits and preconceptions which have later on to be bodily displaced or rooted up in order to secure a proper comprehension of the subject?" (p. 388). Dewey's general allusion to "factors not valid in a strictly demonstrative sense" and "extra-logical factor" point to diagrams as one example but also suggest considering other concrete representations and their spatio-graphical properties. We now tie back these considerations to the conceptions of figure and the secondary geometry curriculum.

2.4.4. Diagrams and Other Representations, and the Conceptions of Figure

If the last section can be summarized, our discussion of the role of diagrams (and, we add, other concrete representations of figures) in geometry seems thus to leave us with the following observations:

(1) Diagrams have played a crucial role in the historical construction of geometric knowledge, scaffolding this construction. Diagrams have made some properties transparent while others were being investigated, and have then supported the bringing of such transparency to question before retreating to the back to become mere illustrations of abstract concepts.
(2) The use of diagrams in communication can be plagued by problems: not only can they lose information, but they can also add unintended information, and they can support multiple meanings with single signs.
(3) A continuity between the elementary and secondary geometry, the latter building on earlier geometric knowledge, may require not the doing away with concrete representations of figure but rather the management of the use of those concrete representations so as to encourage mathematization.
(4) Such management of concrete representations might be part of mathematization itself, as when the more incidental drawings that accompany thinking (Roth & Maheux, 2015) are turned into ready-made diagrams that can be used as models (Azzouni, 2013).

2.5. A Modeling Perspective in the Study of Figures

We are proposing to conceptualize the study of geometry in secondary schools as a process of coming to know geometric figures as mathematical models of the experiential world. Please note that by *experiential world* we don't mean just the real world advocated by proponents of more applied mathematics in the curriculum; rather we mean the world of students' experiences with all sorts of mathematical representations (including, in particular, experiences navigating the space, building replicas, drawing and transacting shapes, sizing up objects, etc.). We mean *modeling* in a mathematical sense; that is, the development of accounts endowed with an internal grammar or mechanism that permits their user to produce (e.g., via some kind of operation with elements of the model) *a priori* predictions about the phenomenon being modeled (reasoning, as Netz, 1998a, would have it, even if not logic). The phenomenon can then be inspected so as to check the predictions. While that describes what a model is supposed to do, the gerund *modeling* points to the recurrent generation of models that improve upon the performance of earlier models, for example by reducing the error of the predictions.

In general, by a modeling approach we are talking about two systems, both of which could be mediated by representations. The simplest way of thinking about that is to imagine one system being a real world phenomenon (e.g., a car wheel that could roll on a flat patio) and the other system being a geometric system of representation (e.g., the diagram of a circle tangent to a line). There will be empirical questions about the real world system (e.g., if a spot on the rubber surface of the wheel is painted with fresh red paint and the wheel is set in motion, how far from the spot where the motion started would the white paint mark the floor?). A modeling approach would discourage answering the question empirically (e.g., "let's roll the wheel and see where the white mark turns up, then we can measure the distance") in favor of creating a mathematical model that permits one to predict an answer. The circle, as a model of the wheel, allows for the development of some concepts and entailments with which to make such a prediction. The process of developing successive such models can contain the development of new geometric ideas (e.g., arc, arc length) and conjectures about those. One could imagine an initial, incorrect prediction based solely on the concepts of circle and distance, expecting that the red mark will be at a distance equal to that between the red mark on the wheel and the initial point of contact as shown by Figure 2.1. Even this incorrect prediction requires some attention to the relative sizes of the wheel and the circle and the assumption that distances are proportional to radii.

The experiment could then be simulated, either with an actual wheel and a red mark on it, or with a representation that preserves some of its physical characteristics (e.g., the cap of a medicine tub rolling on a sheet of paper placed on a table). The prediction could be checked against the experimental measures and its error assessed. In the case of the wheel, that might lead to attempts to a better estimation by using not just the distance between the two endpoints but the sum of lengths of shorter segments, as in Figure 2.2. And so on. The concepts of *arc* and *arc length* could emerge from the attempts to mitigate the experimental error observed when the problem is modeled using the distance between the points.

In the modeling process, elements of geometric knowledge could be developed. Questions about the problem such as, how can one expect that the new model will reduce the error? or, does it matter how to choose the intermediate point?, could lead to the statement of the triangle inequality. While the work on the diagram could initially include some concrete activities such as measuring or counting, the modeling approach could encourage students to use assumptions and plausible reasoning (and at some point also logically necessary inference) along with those measures to anticipate answers to questions. Diagrams then can become a key element in allowing earlier conceptions of figure to evolve into

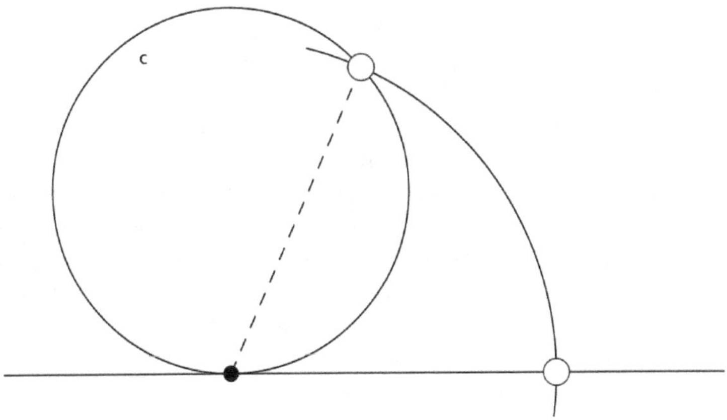

FIGURE 2.1 First approximation at locating the white mark on the floor

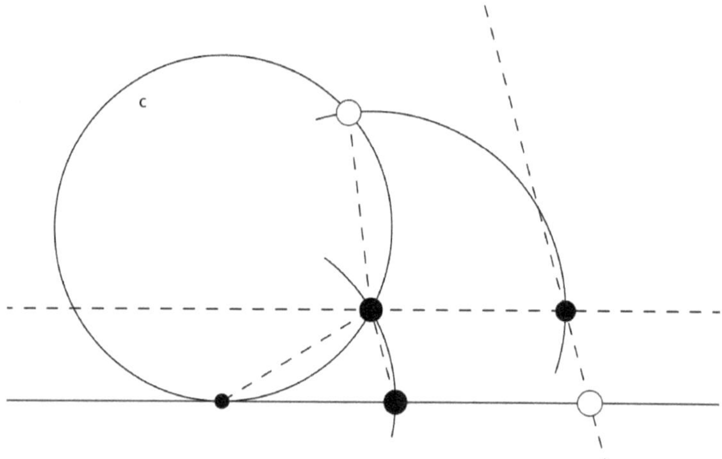

FIGURE 2.2 Second approximation at locating the white mark on the floor

more sophisticated conceptions of figure via a modeling approach. Note, in particular, that while some of this modeling will already have involved diagram work (e.g., in questions that relate to the microspace), other modeling activity will engage diagrams as modeling language (e.g., the case of the wheel above, which belongs in the mesospace, involves construction in the microspace as modeling language). A key question to ask is: How are students to interact with diagrams as they engage in modeling? To flesh that answer out we come back to the four conceptions of figure and briefly sketch how a modeling approach could make those conceptions evolve.

2.5.1. The Modeling Approach and the Four Conceptions of Figure

We sketch connections between the modeling approach and the conceptions of figure here and return to them in Chapter 5. Consider first the conception that we have called *figure as navigation of the macrospace*. We can expect these navigations to rely on at least four kinds of concrete representations, none of which is yet a geometric diagram. These representations include (1) embodied representations, including what it feels like or looks like to walk or run in an environment, (2) iconic representations such as those indicating *play area* or *park* and indexical representations such as those of traffic signs for curves or traffic direction, (3) maps such as city maps or road maps, including internet maps and GIS and (4) representations embedded in the functioning of vehicles with which people navigate space, including the steering controls and wheels of carts, cars, and bikes, sails of boats and windsurf boards, and the viewing devices of vehicles (e.g., windshield, rearview mirrors and camera, etc.). The phenomena to be modeled consists of navigations of space that might use those representation systems and the modeling work will include the development of geometric concepts and propositions associated with the prediction and control of phenomena in that realm. Simple examples could be the activity of describing and eventually making scale drawings of spaces in actual buildings, such as the layout of rooms in a family home, the activity to provide instructions for a robot to simulate daily activities in a home, or the activity to create a scaled replica of a building (such as a house). Each of those activities can engage secondary students with the macrospace, in which they have some initial means of operating via the conception of figure as navigation in the macrospace. As can be noted, the modeling of those elements of the macrospace calls for the creation of new representations (e.g., descriptions, instructions, maps, replicas) that might summon, at least initially, elements of a different conception (e.g., descriptions or constructions in the microspace).

Take for example the case of asking students to provide instructions for a robot to follow as it enters a house, goes to a bedroom, and picks up a book from a desk. One could expect students to initially use language that describes what they would do but that lacks quantitative precision on matters such as how long to walk forward, how much to turn, how high to raise the leg, etc. Simulating the robot's movements could reveal those imprecisions as well as call for considerations of measurement units (steps vs. meters), and so on. In general one could expect that an activity like this one might yield a more sophisticated conception of figure as navigation of the macrospace, in which the navigation can be controlled by transformations and described using various kinds of measurements. Further we can expect this activity to yield diagrammatic representations of navigations that have geometric properties (e.g., traversals of equal space being represented with segments of equal length) that could then be the source of new questioning (e.g., a toy was left on the hallway floor, about three steps from the bedroom door, where in the map could the toy be?).

The second conception we identified is *figure as capture of object in the mesospace*. We can expect students to have means of relating to large objects through gaze (e.g., standing back to seize an object visually), touch and grasp (e.g., putting arms around it to pick it up), language (e.g., saying what it looks like), and bodily movement (e.g., walking around it, climbing it). The task for the secondary geometry teacher is to create modeling activities that summon those representations and help them evolve to more sophisticated conceptions of figure. A good example is the activity proposed by Berthelot and Salin (1998), in which students are asked to identify the exact location on the floor map of a room where a three dimensional object (a bench, in their case) will be transported. As the researchers describe, students have been observed not controlling the angles when drawing the new location of the object; if they could simulate the activity, for example by bringing the object to the drawn location and outlining the space actually occupied by its base they might be able to contrast it with what they drew. Activities of this sort could lead to an understanding of geometric properties that are necessitated by others (e.g., if three angles in a quadrilateral are right angles, the fourth one is right as well).

The third conception we identified is *figure as construction of small objects in the microspace* and concerns in particular young children's drawings and building activities. The register for this conception contains embodied or instrumented representations of geometric entities such as line or rectangle (e.g., a line is a steady trace of a pencil or what remains when one traces the border of a ruler). Activities such as those that have been common in the study of geometry using dynamic geometry could illustrate well what a modeling approach could look like at this level. Students might be asked to construct a given figure with a set of tools (e.g., dynamic geometry software), and they might engage in such construction using only visual controls. They could then explore whether their construction can be "messed up" or is otherwise "unmessuppable" (Hoyles and Jones, 1998), and use the goal of making it "unmessuppable" to then redo the construction using some properties.

The fourth conception we identified is that of *description and manipulation of small objects in the microspace* and concerns in particular young children's knowledge of the names of two and three dimensional figures and their recognition of accomplished transformations (rotation, dilation, etc.) on figures. The register of this conception, unlike that of the first two, already includes geometric figures; the goal of secondary activities should be in part to make that conception more sophisticated by inducing more analytic thinking about the description of figures. Adaptations of the "guess who" game, such as the activity described by Herbst et al. (2005) can support moving from holistic ways of relating to geometric diagrams (e.g., being able to associate a prototypical diagram of an isosceles trapezoid to its name *isosceles trapezoid*) to more analytic ones (e.g., being able to decompose an isosceles trapezoid into a system of questions about single features, such as whether two sides are parallel). As Herbst et al. (2005) show, activities like that one can lead students to infer properties that are necessitated by those already discovered (e.g., if two sides are parallel, the side connecting them makes supplementary interior angles with them). Note that the notion of modeling here is one of articulating systems of properties to account for figures as wholes.

2.5.2. The Modeling Approach and Informal Proof

As noted above, modeling involves more than representing—or to be more precise it calls for a particular kind of representation, one that is operational: A model can be run, its grammar or set of rules can be set in motion, figuratively speaking, so as to make the model calculate or produce information that was not apparent in the initial representation. In the case of using a circle to model the wheel with the red mark, it is clear that the circle not only resembles the wheel and can depict a smaller version of it on paper, but also that the properties of a circle (its constant radius, its unique center, etc.) permit the modeler to operate the thought experiment of rolling the wheel. Note in particular that notions such as arc length do not need to be explicit yet, and still the basic geometric properties of the circle allow the user to operate the diagrammatic representation. This is what we surmise Netz

(1998a) meant by his allusion to diagrams as enabling reasoning if not logic: Our use of modeling builds on Netz's (1998a) idea by suggesting that representations (including diagrams) can model the experiential world of shape and space, not only by depicting them but also, crucially, by enabling reasoning (if not logic, but, at some point or with some representations, also logic). This takes us to considering the role of proof in this modeling approach.

The notion of mathematical proof has a historical and cultural connection with secondary geometry in many countries of the world. In the United States, for example, despite persistent calls for reasoning and proof to be embedded throughout the elementary and secondary curriculum, the high school geometry course has traditionally been the one place where students are consistently expected to learn to do proofs—even though the type of proof commonly found in that course has often been subject to criticism (Herbst, 2002a, 2002b). In proposing a modeling approach we are also attempting a revamp of the role of proof in secondary geometry, albeit one that allows more room for what Lakatos (1976) called *informal proof*. Lakatos' (1976) case study of the historical development of Euler's theorem about polyhedra argues for a dialectic of proofs and refutations as the engine for mathematical discovery; the attempts to prove a conjecture lead to refinements of the conjecture, development of new ideas, even to development of new methods of proof. These proofs are not always or even usually formal, as at various moments they may rely on ideas that have not been formally defined; instead, the informal proofs lead to the development of more precise definitions. This is particularly important when we consider that the modeling of the students' experience with shape and space might be done in school not necessarily with the elements of a geometric theory (e.g., Hilbert's postulates and basic notions) but rather with a system of representation made up of primitives that could be diagrammatic or enactive. We contend that a notion of informal proof might still be available to make those models operational. To make this point a bit clearer we elaborate a bit on the notions of proof that can describe students' experience in school.

In their attempt to address calls for a definition of proof that could help describe mathematical practice in schools, Herbst and Balacheff (2009) proposed that our field needs to relate to proof and proving at least at three levels. A first level could be labeled *contractual* and is related to the general characteristics of the knowledge at stake in a course of studies: Proof may be part of what students are expected to (learn to) do in a course, proof may be described as that which is needed for a proposition to be true, or defined in general as an argument that shows that the conclusion is necessitated by the premises, or contrasted to experimentation as a way of knowing. This contractual notion of proof may be expected ("can you prove that?"), though it does not provide specific means for producing or evaluating a particular proof. A second level could be labeled *situational*, and related in particular to smaller conceptual fields where specific instructional situation, such as "doing proofs" (Herbst and Brach, 2006) or "proving trigonometric identities" (Rosenberg, 1958) may be available to frame the specific work of producing and evaluating a proof. In "doing proofs", for example, it is common to reduce figures to triangles and to operationalize a proof of the congruence of angles or segments in terms of showing they are corresponding parts of congruent triangles. The situation may have norms that regulate specifically what a proof consists of and what counts as each one of its components. For example, in "doing proofs," a proof is expected to be a sequence of statements about generic diagrams, justified by reasons that can be the given, definitions, postulates, or known theorems; hence the conceptual field where such notion of proof can operate needs to be already populated by clear and non-negotiable definitions and at least some postulates (e.g., the postulates for the congruence of triangles; see Herbst et al., 2009). In the case of trigonometric identities, the claim that two expressions are identical has to be warranted on chains of equivalent expressions that might each be validated, though these are not always explicitly noted, by algebraic manipulation, definitions of trigonometric functions, or known trigonometric identities. We note in particular that while many (perhaps all?) trigonometric identities *could* be proven by reference to a geometric diagram, such a proof would be outside the norms that characterize the instructional situation "proving trigonometric identities". The point is that regardless of what may be justifiable on purely mathematical grounds, specific domains, specific situations in school, contain or are characterized by particular sets of operational representations that permit only a subset of the kind of operations (e.g., algebraic, logical) that one associates with canonical mathematical models. Much of the work that students do, however, especially in their experiences with space and shape, is not immediately representable in a system that has a natural way of calculating.

The interplay between the contractual and situational notions of proof can be seen as a counterpoint; while the contractual notion of proof calls for proving as an activity that fulfills a function in regard to knowledge, the situational notion of proof calls for a product (a proof) with specific (that some might describe as formal) characteristics. This counterpoint motivates a third notion of proof, that Herbst and Balacheff (2009) label *metaphorical*, and that relies on the drive and motivation of the former (e.g., mathematical propositions are true if a proof can be shown to exist) and formal guidance from the latter (e.g., statements need to be justified, hence definitions need to be shaped and postulates need to be expressed). Thus the drive to prove propositions in conceptual fields that have not been charted with definitions and postulates could lead, by metaphorical projection[3] (Black, 1962) from fields that have been so charted, to at least provisionally structure such fields. Herbst and Balacheff (2009) show how such metaphorical notion of proof can be relied upon to organize the mathematization of fields of experience[4] that have not yet been so organized in terms of definitions and postulates; they illustrate this using records of a geometry class's work on claims about area at a time when that conceptual field was (for the class) constituted only by the formulas for areas of some basic figures.

The reader may wonder how the above discussion is related to the modeling approach. When one field of experience with shape and space (e.g., going from one place to another in a city) is modeled with resources from another conception of figure (e.g., construction of objects in the microspace, as when one makes a map on paper), claims about the former (e.g., which path will take less time to traverse) might also be represented with resources from the latter (e.g., sequences of segments drawn on paper). The modeling approach would suggest that the grammar of the modeling system (e.g., the map drawn on paper) would enable the validation of those claims by way of running the model. The ways in which the model can be run may vary: A basic one could take the modeling system as merely an icon or a scaled replica of the experience being modeled, as when segments in the map are actually measured with a ruler, and reasoning enters occasionally, such as when a difference in the length of two drawn paths is taken as a difference in the duration of two traversals of space. More sophisticated ways to run such a model could take progressive account of what the elements of the representation stand for (e.g., segments are city blocks while endpoints are street crossings) and validate claims on the basis of combining those representations logically along with postulated assumptions about them (e.g., assume that crossing an avenue takes three times as long as crossing a street). The counterpoint between a contractual notion of proof, that is, the currency of an expectation that claims need to be validated by proving, and earlier, situational experiences with specific inferencing, that is the capacity to reason through specific types of claims (e.g., in solving equations or in calculating with multi-digit numbers), can be used to construct (by way of metaphorical projection; Black, 1962) informal proofs to validate claims by reasoning about the representations in the modeling system.

2.6. Conclusion

Clearly, the paragraphs above only hint at the possibilities of modeling as a way of improving those four conceptions of figure. Chapter 5 comes back to them after Chapters 3 and 4 deal with two important aspects of the work of secondary geometry education: The students' capacities to think geometrically and to learn geometry and the teachers' knowledge and practices in geometry instruction. As regards the objectives of Chapter 2, the following serves as a conclusion.

Secondary geometry provides the means to predict what actions will work and explain why those actions may not have worked in students' interactions with representations of figures. While spatio-graphical properties are a large set of predicates that describe (in some cases, just quote), from an observer's perspective, all the claims about figures that can be made in activities of navigation, capture, description, and construction, theoretical properties are a smaller set of predicates that prescribe or stipulate those claims that will serve to explain, control, and predict the outcomes of those activities, given some constraints (e.g., descriptions of hard facts about the material world such as assuming that the edge of a ruler is a straight line).

More synthetically: If the study of geometry in secondary school can produce any new learning in students, this learning is likely to be about the control structures that regulate the interaction between the agent and representations of figure within and among older conceptions of figure. Those control structures will be, in particular, the systems of informal proofs that can regulate the activities of rationally predicting and reasonably explaining the way those conceptions of figure work.

Notes

1 We use *replicas* to refer to artifacts such as scale models. We thus preserve the word *model* for its use in naming what is accomplished by the practices of modeling experiences with shape and space.
2 Brunner-Traut, 1974/1919, seems to have related this contrast to a contrast between additive and integrative perceptions of the object being represented; see Sörbom, 1994.
3 Black's (1962) notion of metaphorical projection holds that in a metaphor one posits an association between two systems, the structure of one of the 0s00y0stems is projected or imposed onto the other system, in much the same ways as a morphism in mathematics can project structures from one set to another.
4 We are using this expression in a relaxed sense to allude to experiences with shape and space that might be filed under any of the four conceptions identified above. We believe our use is not incompatible with that of Mariotti and her colleagues (see Mariotti et al., 1997).

3

STUDENTS' THINKING AND LEARNING IN GEOMETRY

3.1. Introduction

This chapter adds a consideration of students' thinking and learning to our prior considerations of history and epistemology. In chapter 2, we propose that the study of geometry in secondary schools may be thought of as a process of coming to know figures as mathematical models of the experiential world. We propose four conceptions of geometric figure as organizing younger students' experiences with shape and with the various spaces that they participate in: (1) figure as navigation of the macrospace, (2) figure as capture of an object in the mesospace, (3) figure as construction of small objects in the microspace, and (4) figure as descriptions and manipulations of small objects in the microspace. From an epistemological perspective each of these conceptions can be characterized by identifying the set of problems it solves (P), the set of operators used to solve those problems (R), the semiotic registers used to represent the work (L), and the set of controls used to regulate operations and validate solutions (Σ) (Balacheff & Gaudin, 2010). It is important to keep that perspective in mind as we look more closely into the research on students' thinking and learning.

We also indicated that secondary geometry provides the means to predict what actions will work and explain why some actions may not have worked in students' interactions with representations of figures. Thus the study of geometry in secondary schools could enrich the control structures that regulate the interaction between the agent and representations of figure within and among those older conceptions of figure. From a cognitive perspective we explore how adolescents solve geometric problems, in particular with respect to the modeling approach and the use of informal proofs discussed in the previous chapter. In inspecting such problem solving we synthesize what the cognitive literature says about the different representations and controls that adolescents use when they work in problems that belong in the various spaces in which they interact.

Note that thus far we have used the word *representation* in this book to refer to a material artifact that stands in for, evokes, or points to a mathematical concept. The cognitive literature has used the word *representation* in ways that make room for the questions asked above—namely of the relationship between adolescents' thinking in geometry and the representations of figure they interact with in each of the conceptions in which they come in contact with geometric figures. For example, Mesquita (1998, p. 184) observes that *representation* can refer to both internal (mental) and to external (concrete) representations, and asserts that an external representation of a geometrical figure does not by itself enable students to explore its properties. Rather, it is inner representations, i.e., visual imagery, that enable (but possibly also restrict) students' thinking (Dreyfus, 1995, p. 6).

We have seen that geometric diagrams play key roles in problem solving processes. Koedinger and Anderson (1990) made observations of how experts solve geometric proof problems and found that in their proof planning they tend to focus on only essential steps of reasoning, working forward to conclusions, and making use of diagrams (external representations) very effectively. In particular, they quickly construct a network of related diagrammatic configurations to effectively plan and solve geometric proof problems. But do adolescents have such mental abilities and capacities when they solve problems in geometry? Starting from this question, in section 3.2 we discuss the following subquestions: How can students benefit from using representations of figures to model their experiences with space and shape? How can their means of control use the increased explicitness and operationalization that one might expect from mathematical models? We review below how students create internal representations out of their experiences with geometric figures; in particular we return to Fischbein's (1993) notion of *figural concept* (briefly mentioned in Chapters 1 and 2) and consider Tall and Vinner's (1981) notions of *concept image* and *concept definition*. We also present examples to illustrate how such theories can be used to describe and explain students' misconceptions and difficulties in geometric reasoning with figures. Such discussion is particularly essential to understand how modeling tasks engage students' cognitive processes and practices of informal proving with geometric figures.

In section 3.3 we address the progression of students' geometric thinking in the learning of geometry. In general, by "learning" we mean how students progress from less to more sophisticated cognition of shape and space. Within that space of questions, we may identify at least three subquestions: (a) By which mechanisms or processes do cognitive structures or patterns of interaction evolve? (b) Through which developmental paths does cognition or interaction pass? and (c) In the context of which environments does cognition or interaction change? In this chapter, we address the first of these questions briefly, and then provide a substantial discussion of the second. We pay particular attention to the van Hiele theory, in which individual development is described as a passage through distinct levels of geometrical thinking. Various studies have used the van Hiele theory, which was foundational for research on the learning of geometry even if it is now more of a historical artifact. We describe more recent attempts to improve as well as move away from the original van Hiele levels. We also discuss various ways of capturing changes of geometrical thinking over time, including a discussion of students' thinking with respect to Dynamic Geometry Software (DGS) in section 3.4. In section 3.5 we describe the construct *learning trajectory*, which includes "the learning goal, the learning activities, and the thinking and learning in which the students might engage" (Simon, 1995, p. 133) in order to examine how the notion of such a progression can help the field move beyond what the van Hiele model can offer (Clements et al., 2004; Clements & Sarama, 2014). Chapters 4 and 5 include consideration of the task environments in which learning happens.

3.2. Conceptions of Figure and Students' Cognition

3.2.1. Connecting Conceptions of Figure and Modeling to Students' Cognition

In this section we review research on students' cognition and elaborate on what the research says about cognitive process in terms of the four conceptions of geometrical figures. The modeling approach proposed in Chapter 2 relates two different systems: An experiential world of shape and space, and a representation system (possibly, but not necessarily, including diagrams) with which the first one is modeled, not only by depicting the experiential world but also by enabling reasoning that predicts and explains experiential phenomena. As we have noted, informal proofs play a prominent role in such reasoning. But proving in geometry is not an easy task for many students, many of whom find it difficult to distinguish between measurement and deductive proof (Clements & Battista, 1992; Battista, 2007). Chazan (1993) found that adolescents have confusing views and beliefs about empirical evidence and proofs: For example, students often mistakenly think that empirical evidence can be used to guarantee that a geometrical statement is true in general; at the same time, they may mistakenly regard a proof as nothing more than evidence that a claim is true for the single case depicted by a diagram.

In Japan, geometry is the main context for teaching and learning proof (Fujita & Jones, 2014), but national surveys repeatedly demonstrate that proving in geometry is one of the areas that students find particularly difficult. They often struggle to plan their proofs from given information and diagrams, to bridge assumptions, intermediate propositions and conclusions, to use congruent triangles in order to deduce conclusions, and so on (see also Miyazaki et al., 2015). Geometric constructions with straightedge and compass can also have a problematic relationship to proving. Schoenfeld (1989) reported how high school students who successfully completed a full-year geometry course failed to see a connection between a proof problem that they had studied and a related geometric construction, operating instead as if proof and construction tasks were totally distinct. Tam and Chen (2012) have similarly noted that students often find it difficult to link construction procedures to their underlying geometric properties.

We argue that by taking a modeling approach we might be able to support students in transitioning away from the simple perception and empiricism with which they control their responses to mathematical questions and towards the use of reasoning to predict an answer, that is by engaging in informal proving. As noted before, we use the phrase *informal proof* in the sense of Lakatos (1976), to describe activity in which students rely on ideas that may not have been formally defined and use them to make naive conjectures, reduce the problem to simpler problems articulated deductively, find counter-examples, refine the conjecture, revise the definitions and reduction, etc. The proposal of a modeling approach, in which the conceptions of geometrical figures and the notion of informal proof are used to model students' experiences with shape and space, suggests the need to investigate relationships between adolescents' geometrical thinking and their interactions with representations of geometrical figures.

The literature from cognitive psychology supports the notion that such a modeling approach can be effective in engaging students in proof and proving. Stylianides and Stylianides (2008) reviewed two cognitive theories that purport to describe how young learners engage in deductive reasoning. In one of those, the mental model theory proposed originally by Johnson-Laird (1983), Bauer and Johnson-Laird (1993) claim that "individuals reason by (a) constructing a model, or models, based on the information in the premises and background knowledge, (b) formulating a conclusion that is true of the model and subject to other constraints, such as parsimony, and (c) searching for alternative models in which the conclusion does not hold" (p. 372). The researchers noted that the use of a diagram allows learners to keep track of alternative possibilities, and this dramatically improved learners' reasoning performance.

For example, in a Japanese grade 8 textbook, students are presented with the problem "In an amusement park ride like the one shown below, the platform where people ride always remains level. Let's think about why" (Fujii & Matano, 2012, p. 133; this is accompanied by a photograph that can be seen at http://moriumi.web.fc2.com/zhitachi-rejarando.html). This problem summons students' experience in the macrospace (e.g., riding a moving surface without falling), but the ride itself (from an observer's perspective) can be described as an object in the mesospace. In order to investigate the properties of parallelograms, the problem can be modeled by figures in the microspace, such as drawings on paper or on a computer screen, or physical models. Once modeled, the problem can be mediated and controlled by the conception of description of small figures in the microspace (see Figure 3.1). Students may examine what representations can be used, what shapes can be formed (e.g., parallelograms), and what definitions of the formed shapes can be referred to, in coming to recognize the need to prove a general statement such as "If two pairs of opposite sides of a quadrilateral are parallel and equal in length, then the figure is a parallelogram" which can in turn be interpreted in terms of the physical situation in the mesospace to explain why "the platform where people ride always remains level".

Another example, studied by Jahnke and Wambach (2013), concerns the yearly path of the sun under the paradigm that "the earth is a sphere and located in the

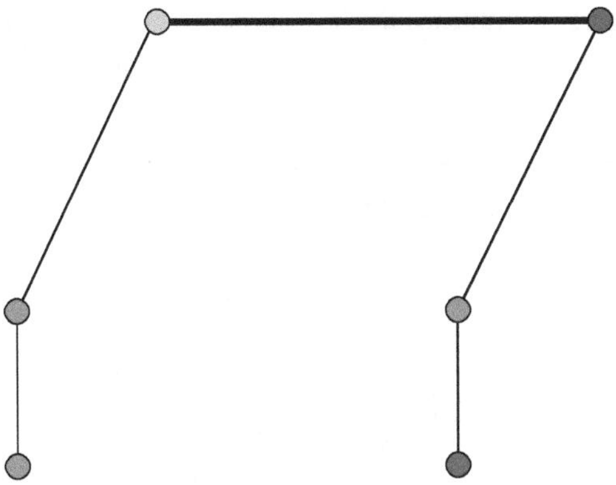

FIGURE 3.1 Modeling a park ride

center of the universe" (p. 473), the predominant cosmological system until the seventeenth century. From the perspective of the conceptions of geometric figures, the problem evokes students' conception of figure as navigation of the macrospace, in that students have direct experience of seeing the sun moving in the sky. Jahnke and Wambach recount the history of how Hipparchus of Nicaea (c. 190 – c. 120 BCE) constructed a model based on this first hypothesis ("The center of this circle is located in the center of the earth") and subsequently modified his model and hypothesis by referring to observed empirical data that conflicted with the predictions of his model. Here, the problem is modeled by drawn diagrams, that is by using the conceptions of figure as construction or description of small objects in the microspace. Those figures enable alternative models so that Hipparchus could eventually derive a new hypothesis, namely, that the center of the sun's circular orbit is not located in the center of the earth. A theorem —"There is exactly one position in which the sun is the farthest from the earth and exactly one in which it is nearest to it"—was consequently deduced; this theorem was then proved by reference to abstract properties of triangles and circles.

It is possible, however, that students' reasoning might lead them to error. In terms of Chapter 2's discussion of seeing-as, we could ask whether seeing-as suggests spatio-graphical properties that might not be theoretically provable. Indeed the thinking of many students is influenced by visual representations that may be incorrect. Kospentaris et al. (2011) reported that even 18-year-old university students have difficulties employing strategic approaches such as deductive reasoning to judge equivalence of two or more shapes' areas, tending instead to employ intuitive judgments such as "the longer the sides of the figure, the greater its area" (p. 123), or visual judgments influenced by what the diagrams look like. Biber et al. (2013) found that 14-year-old students could not utilize properties of angles in lines and polygons in order to find sizes of angles. They also report that the main cause of this difficulty might be students' perceptions of visual aspects of drawings of geometrical figures (p. 58). In the next section we further scrutinize students' cognitive difficulties in problem solving with geometrical figures, in particular focusing on students' thinking around figural and conceptual aspects of geometric figures.

3.2.2. Geometric Thinking as Interactions Between Figural and Conceptual Aspects

3.2.2.1. Fischbein's Figural Concepts *of Geometric Figures*

We have proposed that geometry in secondary schools provides a means for predicting what actions will be effective, and to explain why some actions would not work, in the context of handling problems that belong in the four conceptions of figure described in Chapter 2. From the point of view of student's thinking and learning, we need to know what cognitive functions describe a learner's process of perceiving geometrical figures, and how students can develop more sophisticated control systems which regulate various operations when solving problems. Let us take the variation of the "guess who" game mentioned in Chapter 2, in which Student A thinks of a figure and Student B asks a sequence of yes or no questions in order to determine what figure A was thinking of. Suppose student A imagines a rectangle, and student B asks "Is it a parallelogram?" What responses can we predict that student A might offer? If student A responds "No", then what, on the basis of that response, can we infer is student A's definition of a rectangle, or what cognitive process would have been involved in reaching this answer?

One possible approach for considering matters of this kind is to elucidate how the visual representation allows operations of thought that are supported by the manipulation of the representation language. Fischbein (1993) proposed the notion of *figural concept* to address the fact that while a geometric figure (e.g., a parallelogram) could be described as having properties that can be inferred logically from its definition and other assumptions of a theory (i.e., Euclidean geometry), some other properties can be inferred from the mental manipulation of visual representations of the figure (ibid., p. 141). From this point of view, in the example proposed above, where student A denies that a rectangle is a parallelogram, we can ask what controls regulate this decision. Let us consider this by focusing next on the roles of definitions of geometric figures.

3.2.2.2. The Roles of Definitions of Geometric Figures

Duval (2002) argues that learners need to "[become] aware of the discrepancy between a valid reasoning and a non-valid reasoning" in geometry (p. 63). In the modeling approach we propose, the deductive reasoning involved in informal proofs is one of the forms of valid reasoning used to predict, control, and explain those actions. We do not expect students to make deductions starting from given, formally defined ideas. Rather, we take the iterative development of more precise definitions of geometrical figures to be one of the key learning processes. The geometric notion of movement is a historical example: The notions of movement and superimposition aided inference and helped establish geometric truths long before any theory of geometric transformations could provide the precise analytic language required for logical derivations.

In the notion of *figural concept*, conceptual definitions of geometric figures also play key roles in students' understanding. Indeed, the ways in which students define mathematical concepts are deeply related to their understanding of geometric figures: In the learning of geometry, just knowing the definition of a geometric figure does not imply a full understanding of such figures (Zazkis & Leikin, 2008; de Villiers, 1998; 2010). Levenson et al. (2011) state that geometric figures are "concepts, abstract ideas derived from formal definitions" (p. 8). This definition of figure suggests that a full understanding of a figure might include being able to spell out the set of statements about the figure that can be logically derived from a definition. But also as discussed in Chapter 2, geometric figures also have visual representations in the form of diagrams. These diagrams are not solely physical artifacts; the word representation identifies them as instances of semiosis encoded in a system of meaning making that also affords inferences (though possibly not, or not only, the inferences of logical necessity).

The terms *concept image* and *concept definition* are useful to describe the control structures students use in their thinking about geometric problems. Those terms were introduced by Vinner and Hershkowitz (1980) in the context of the learning of some simple geometric concepts, and developed by Tall and Vinner (1981) in the context of more sophisticated mathematical ideas of limits and continuity. They defined a *concept definition* as "a form of words used to specify that concept" and *concept image* as "the total cognitive structure that is associated with the concept, which includes all the mental pictures and associated properties and process" (Tall & Vinner, 1981, p. 152). Thus, a parallelogram (within Euclidean geometry) could be defined as a quadrilateral which has two pairs of parallel sides (concept definition), but the parallelogram is also associated with some geometric images, in which the parallelogram could be seen as having different side lengths and angle sizes between concurrent sides. Connecting to our consideration of conceptions of geometric figure, at least as it refers to the microspace conceptions, we can say that students' control structures include conceptual and figural aspects of geometric representations that can be described in terms of concept-definition and concept-images.

The case of reasoning about quadrilaterals provides an interesting example of how research has documented different ways in which students deal with defining or recognizing a figure, which informs our knowledge of students' control structures. Michael de Villiers (1994) has noted that while both hierarchical[1] and partitional[2] definitions for quadrilaterals are mathematically correct, hierarchical definitions are more conventional because they are economical: Not only do they rely on a smaller number of stipulations, but also they allow figures to inherit truths proved about other categories of figures that include them.[3] Hyman Bass agrees with this position when reflecting on how mathematicians think: "mathematicians generally favor inclusive definitions (a square is a rectangle; a parallelogram is a trapezoid; etc.). The main rationale is that theorems about the general concept apply ipso facto to the special cases" (personal communication to P. Herbst, August 30, 2015).[4] Yet the representations of quadrilaterals (e.g., diagrams included with definitions) tend to also include spatio-graphical properties that might support reading partitional definitions (e.g., the usual representation of a parallelogram might support reading in an implied condition that consecutive sides make non-right angles). Here we ask how students respond to problems that require them to define or classify quadrilaterals. To answer this, it is reasonable to expect that students have a concept image and concept definition for geometric figures and that those resources support their engagement in geometric thinking, but also that students' thinking about geometric figures is rather complex and their use of those diverse resources in geometric problem solving (in particular, classifying figures) often contains a range of errors (e.g., de Villiers, 1998). In the next section we further illustrate how cognitive research has grounded the notion that the same figure may be addressed by different conceptions among secondary school students, in particular by the deployment of different control structures.

3.2.2.3. Different Conceptions of Geometric Figures in Geometric Thinking

As we have seen in a problem such as the "guess who" game, problems in which students have to define, describe, classify, or compare plane figures are common in secondary school and appeal to a prior conception of figure as description and manipulation of small objects in the microspace. While those problems might be contexts for the use of deductive controls as students reason about figures, such problems might also participate in different conceptions, particularly in conceptions where the controls are different.

For example, suppose we ask a student who has learned the hierarchical definitions of quadrilaterals to define a parallelogram. If this student can answer that a parallelogram is a quadrilateral with the opposite sides parallel to each other, then can we assume this student has his concept images of parallelograms hierarchically organized so that the statement may also be applied to special types of parallelograms such as squares or rectangles? And, when we ask whether it is possible to draw a parallelogram whose four vertices are on the circumference of a circle, can we expect that same student to say that it is possible because a rectangle or square can be so inscribed, and they are also parallelograms? However, the majority of students say that it is not possible, or that such a parallelogram does not exist (Koseki, 1987; Fujita 2012). They often deduce "a parallelogram does not fit inside a circle as only two sides will touch," which is controlled by their concept image of parallelogram which include spatio-graphical properties, or perhaps only images of parallelograms that respond better to a partitional definition (i.e., they do not have right angles).

From the point of view of conceptions of geometric figure, it seems that there are at least three kinds of control structures organizing the responses to the question, "What figure is this?" The first control structure is perceptual and relies on prototypes, or figures which include extra spatio-graphical properties, such as a particular orientation; these prototypes are part of the concept image students have. One useful theory to consider students' controls based on limited concept images is the prototype theory (Rosch, 1973). Prototypes play important roles in learners' cognition; within the context of the teaching and learning of geometry, Hershkowitz (1990) identified a prototype phenomenon, described as follows:

> Each concept has one or more prototype examples that are attained first and therefore exist in the concept image of most subjects. The prototype examples were usually the subset of examples that had the "longest" list of attributes – all the critical attributes of the concept and those specific (non-critical) attributes that had strong visual characteristics.
>
> *(p. 82)*

For example, parallelograms are introduced as a quadrilateral with two pairs of parallel sides, but typically a slanted representation is used. This image will stay strongly in a student's mind. Hershkowitz also suggests that this prototype representation of a parallelogram (i.e., a 'slanted rectangle') may also lead students to think that "parallelograms do not have right angles" (Fujita, 2012). Thus, although many students are able to state a correct definition of parallelogram, their *concept image* nevertheless still contains a slanted image for parallelograms, and this concept image conditions their reasoning, resulting in statements such as "parallelograms don't have right angles". As a result, they may not consider a square or rectangle as a parallelogram because they have right angles.

The second control structure is descriptive and is composed of a disorganized set of properties that are true about a figure. As a consequence, some students define figures by listing all the properties they know and do not seem to give careful considerations to how they might entail each other. For example they might say that a rectangle has four right angles and two pairs of parallel sides, or a parallelogram has two sets of parallel lines, diagonally opposite angles are equal, and opposite sides are equal (Herbst et al., 2005).

The third control structure is both hierarchical and economical, and this structure is in play when the act of defining or checking for properties is done using the minimum and most powerful properties first. This is a useful way of investigating properties of shapes. For example, if a statement is true for parallelograms, (e.g., the diagonals of parallelograms intersect at their midpoints) then this will also be true for squares, rectangles and rhombi, and in general this statement does not have to be re-proved for each of these types of parallelogram (although see Endnote 4, this chapter). The research evidence suggests that many students have difficulties mastering this control structure (e.g. Koseki, 1987; de Villiers, 1994; Walcott et al., 2009; Fujita, 2012; Özerem, 2012), though some work suggests that even young children can do it in the context of work with DGS (Battista, 2007; Kaur, 2015; Sinclair & Moss, 2012).

3.2.2.4. Dimensional Deconstruction of Figures

Related to the control structures described in the prior section are Duval's (2005) two opposite modes of visualization: Iconic and non-iconic. These describe different sets of cognitive operators that handle the passage from a "discriminated recognition of forms to the identification of the objects to see" (p. 13, our translation). While humans may rapidly perceive distinct forms, the identification of what geometric object is being represented by a diagram is neither fast nor univocal. Duval (2005) introduces a *mechanism of iconicity,* consistent with the prototype phenomenon noted above, to describe how the resemblance between the recognized form and the prototype of the geometric object support the passage between recognition of forms and identification of objects. *Iconic visualization* is operated by this mechanism of iconicity that, according to Duval (2005), maintains the figure as an object which is identified independently "of any operations that one can do on it" (p. 14). Iconic visualization treats geometric diagrams as other visuals (e.g., photographs, caricatures), establishing holistic correspondences between them and the figures they are meant to represent. In contrast, Duval (2005) defines *non-iconic visualization* as "a sequence of operations that permit us to recognize geometric properties by the impossibility of obtaining some configurations or by the invariance of the configurations obtained" (p. 14). In this mode of visualization, the figure is identified as a "configuration detached contextually from a more complex net or organization" (p. 14) of possible configurations. This non-iconic visualization is the cognitive mechanism that makes possible the realization of the epistemological possibility described in Chapter 2 as "seeing as" (Coliva, 2012). Duval (2005) introduces the notion of *dimensional deconstruction* as a way of explaining how cognition makes possible non-iconic visualization.

Duval (2005) describes the non-iconic visualization by distinguishing interactions with the diagram. The actor might introduce "supplementary traces" (p. 16) or not do so (though they might imagine those supplementary traces). Among those moves that rely on the introduction of supplementary traces, Duval (2005) distinguishes between an *instrumental* visualization that relies on traces that are *auxiliaries,* used to construct a figure with instruments such as straightedge and compass, and a *heuristic* visualization that relies on traces that are *reorganizers,* used to solve a problem and independent of the instruments utilized (e.g., possibly including freehand drawings). Duval notes that the need to make supplementary traces compels the actor to relate to the lower dimensional elements that constitute the figure—for a 2-dimensional figure such as a parallelogram, its lower dimensional elements are sides (1-dimensional) and vertices (0-dimensional). For example, the heuristic move that divides the parallelogram into two triangles by drawing a diagonal compels the actor to notice the vertices through which the diagonal should pass. Duval notes that this kind of noticing, which is done through these heuristic (or *generative*; Herbst, 2004) interactions with the diagram, need not be mediated by the statement of properties or names. That is, the activity of drawing supplementary traces supports dimensional deconstruction without the need for language representations that precede the deconstructed elements of a figure.

This perspective thus presents the possibility that activities with diagrams might be sources of experience in which to develop or otherwise introduce names for objects and statements of properties encountered. Duval proposes that there is a possible evolution in students' conceptions of figures that goes from an iconic visualization to an instrumental (non-iconic) visualization, then to a heuristic (non-iconic) visualization.

However, statements are also crucial in order to learn geometry. As Duval (2005) notes, propositions and theorems, rather than merely the words of a specialized vocabulary, are the carriers of geometric meanings in language. It is possible that though visualization may carry attention to particular objects that make up a figure, the language representations of the relations in which those objects are involved and that contribute to the meaning of a figure may lag behind: Duval (2005) calls this phenomenon a *dimensional hiatus*. For example, the act of drawing the diagonal of a parallelogram calls attention both to the vertices and the segment between them, but they do not necessarily imply that the actor considers the statement *the diagonal of a parallelogram separates a parallelogram into two triangles*. The point is that this discursive representation of the geometric object may need different activities, not necessarily contained in those of construction and exploration of a diagram. The production of statements about figures requires new problems in which the lower-dimension objects visualized are put together into reconstructions of higher-dimension objects. Problems that involve students in proving or arguing claims about a figure can contain that sort of reconstruction.

There is much more to gain from a reading of Duval's (2005) paper than we can fairly represent here, but the above summary suggests a few things that both inform our perspective on cognition and learning and connect with the modeling approach proposed by this book. Duval (2005) contributes to our understanding of the conceptions of figure from the perspective of the individual actor by specifically providing ways to account for operations that the actor may do with representations of a figure, and how these proximal interactions with diagrams support visualization as well as discourse. The different kinds of visualization describe qualitative differences of increasing complexity in students' thinking and can thus be used to describe learning. The description of the construction of geometric knowledge as the departure from iconic visualization through dimensional deconstruction and the build up of geometric meaning by discursive representation locates geometric thinking and learning in students' activities in the world, mediated by diagrams, language, and other signs.

Inasmuch as secondary geometry aims to create intellectual tools for students to model their experiences with shape and space in the context of the four conceptions introduced in Chapter 2, Duval's (2005) contribution helps flesh out how models are to be constructed and what it means for them to be models. The creation of non-iconic visualizations is one crucial aspect of this modeling process, and it can be supported by engaging students in activities of reproducing experiences in shape and space using diagrams or other concrete materials and tools: Making maps and scale models, and reproducing and transforming diagrams, can be used as principles for designing learning activities insofar as they can promote dimensional deconstruction. However, models are not merely representations, they rather are calculable representations; they can be cranked out to produce assertions that are not apparent in the representations themselves. For diagrams (and other models) to do that job, they need to be embedded in tasks that go beyond the making of representations and that problematize such making in order to promote discursive reorganization of lower dimension objects in higher dimension objects. We consider examples of this in Chapter 5.

3.2.2.5. Different Conceptions of 3D Figures

In Chapter 2 we identified the conception of figure as capture of large objects in the mesospace. While younger children's interaction with those large objects may be mostly perceptual and proprioceptive, adolescents may be expected to model those large objects using verbal and diagrammatic representations. In such a context, their capacity to reason about geometric figures may require the engagement of sophisticated control structures. Geometric figures can be used to model such large objects, and one useful way is to draw representations of 3D objects on a piece of paper (Ben-Chaim et al., 1989). There is thinking that happens around the drawing activity, but how do students think about 3D figures with 2D representations? In Chapter 2 we referred to Parzysz (1988) who suggests that drawings or diagrams of three-dimensional figures necessarily lose information to the point of inducing a dilemma in students, between *knowing* and *seeing* the properties of 3D figures. Ben-Chaim et al. (1989) also reported that secondary school students had great difficulties in communicating properties of large 3D objects that were verbally and graphically presented in 2D formats.

To understand the cognitive aspects of the dilemma between *knowing* and *seeing*, we find useful the framework "visual perception and perception-based knowledge representation (VPR)" proposed by Gal and Linchevski (2010). According to them, when we solve geometric problems we involve the processes of *organization, recognition* and *(mental) representation* (p. 166). First, in *organization,* shapes and objects are extracted from the visual scene (e.g., from drawing or diagrams). Then, in *recognition,* those objects and shapes are transformed as meaningful mental objects. Two types of processing are considered: Bottom-up (using information from the sensory physical stimulus for pattern recognition) and top-down (using general knowledge for pattern recognition; p. 170). Finally, recognized objects are represented mentally in various ways (e.g., verbally or visually). This *representation* can be used to make sense of a diagram's meaning, and the diagram can be used to solve geometric problems.

In the process of *organization, recognition,* and *mental representation*, students also have to exercise spatial abilities in order to control their decision-making. Pittalis and Christou (2010, pp. 192–194) observe that various abilities are involved in 3D geometrical thinking, including recognizing the properties of 3D shapes and comparing 3D objects, recognizing and constructing nets, structuring 3D arrays of cubes, calculating the volume and the area of solids, and manipulating different representational models of 3D objects. They argue that the interpretation of 2D representations of 3D figures utilizes two main capabilities: a) recognizing the properties of 3D shapes and comparing 3D objects, and b) manipulating different representational models of 3D objects; both of these capabilities face particular challenges. Furthermore, students' reasoning is also related to spatial abilities such as spatial visualization, spatial orientation and spatial relations (Pittalis and Christou, 2010, p. 195).

For example, in order to extract shapes from given diagrams, students might have to exercise "spatial relations", the ability to mentally rotate a spatial object as a whole quickly and correctly, and "spatial orientation", the ability to remain unconfused by the changing orientations. Pittalis and Christou (2013) also studied cognitive processes of decoding (interpreting the structural elements and geometrical properties) and coding (constructing 2D representations and nets of 3D shapes, and translating from one representational mode to another) of 3D shapes in 2D representations. One of their findings is that in the decoding process, (i.e., in interpreting representations of 3D figures) students utilize two spatial abilities, recognizing the properties of 3D shapes and comparing 3D objects, and manipulating different representational models of 3D objects.

Let us see one example in which students show a range of responses when they solve arguably simple problems. For example, Jones et al. (2012) and Kondo et al. (2014) reported that only 14 percent (N=220) of 15-year-old students could deduce the size of the angle *DEB* in a cube correctly in the figure below (60 degrees) by deducing first that triangle *DEB* is equilateral (see Figure 3.2).

In the case of the problem shown in Figure 3.2, first some geometric information can be extracted from the given representation. In particular it is essential to extract angle DEB from the diagram—an example of the recognition process.

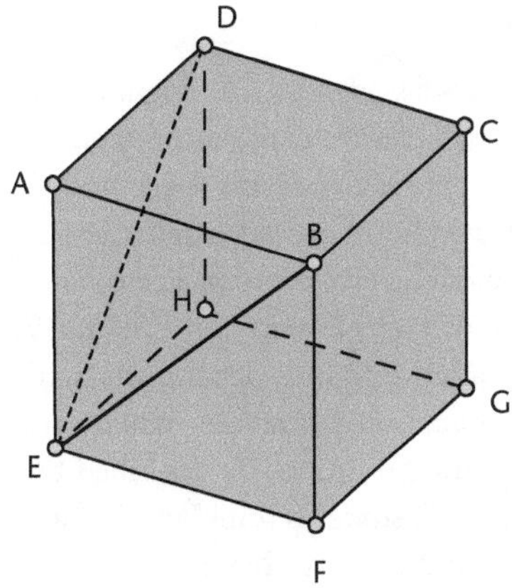

FIGURE 3.2 What is the size of the angle *DEB*?

An adaptation of Kondo et al., 2014

Also, the lengths of $\overline{DE}$, $\overline{EB}$ and $\overline{DB}$ can be deduced from the properties of a cube (e.g., all diagonals of its faces are congruent). If the segment $\overline{DB}$ is considered, then a triangle *DEB* might also be recognized as equilateral (by top-down recognition). As one thinks about this as an equilateral triangle, the target angle size of 60 degrees can be deduced. However, the findings by Jones et al. (2012) and Kondo et al. (2014) suggest that such process is not likely to occur, and students utilize a variety of different operations such as directly measuring angle *DEB* by directly applying a protractor on a representation; drawing a net and then using 45 + 45 = 90; or rotating the perspective and saying triangle *DEB* is an isosceles triangle with an angle of 90 degrees. Other students might be able to answer this question correctly by an operation, for example drawing a segment $\overline{DB}$ and then finding that *DEB* is an equilateral triangle.

These findings indicate that reasoning about 3D shapes with 2D representations is controlled by various factors, in particular by which visual images students extract, indicating that there are different control structures for conceptions of 3D shapes with 2D representations. For example the most naive control structure might be decoding a 3D representation based on visual appearance of a representation. Such a control structure regulates operations such as directly applying a protractor on the representation. The second control structure is to activate an operation to draw/consider a net in order to visualize the angle *DEB*, which does not lead to a correct answer in the problem. The third control structure is to summon a figure influenced by the visual information of the representation, such as a right isosceles triangle, and then deduce the size of the angle based on the properties of an isosceles triangle. The fourth control operates by drawing a segment $\overline{DB}$, recognizing the properties of a cube, and then deducing the relative lengths of $\overline{DE}$, $\overline{EB}$, and $\overline{DB}$.

3.2.3. Conceptions of Geometric Figures from a Cognitive Perspective

This chapter has thus far discussed students' cognition about geometric figures in terms of questions such as: How can students benefit from using representations of figures to model their experiences with space and shape? And, how can students' means of control use the increased explicitness and operationalization that one might expect from mathematical models? For these questions, existing theories (such as those by Fischbein, Hershkowitz, or Tall and Vinner) suggest that internal representations of geometric figures depend on figural and conceptual aspects that cannot be easily separated. For example, when using the control "if a shape does not have a property, then it cannot be described by a name whose definition requires having that property", students think visually and conceptually. When students engage in reasoning and proving they need to engage in a dialectical process (Mariotti, 1995) of reconciling conceptual and figural aspects of shapes (p. 104). At the same time, students' limited control structures might result from conflicts between figural and conceptual aspects. In particular, prototypical examples stay strongly in students' mind and, in order to engage deductive reasoning in geometry, they need to overcome the limits of such internal images of geometric figures. In terms of the thinking process, in particular in terms of making sense of geometric figures represented in various forms, the VPR theory proposed by Gal and Linchevski (2010) suggests that the process of perceiving external representations involves several steps: Extracting relevant information, forming internal images to recognize information, and representing one's thinking in various forms. In this process, visual information again plays key roles, and bottom-up recognition (using information from the sensory physical stimulus for pattern recognition) often leads students to reason incorrectly.

An obvious implication is to consider what teaching interventions would be necessary to promote such shifts in thinking, but before considering this important issue, we discuss the development of geometric thinking (i.e., progression from less to more sophisticated cognition of shape and space).

3.3. Geometric Figures and Students' Learning as Progressive Change in Cognition

We take *learning* to mean *progressive* (as opposed to regressive) *change* in cognition—including not only increase in the items known but also increase in the quality of that knowing. Chapter 2 has introduced the epistemological notion of *conception* which enables a description of what is known in terms of practices, and which affords the possibility to think of learning as changes in the practices with which students interact with their milieu; such learning could be seen as progressive change in the quality of students' performance in a given practice, as well as progressive change in the number of practices in which the student can perform.

There are strong resonances between that way of conceptualizing learning and current approaches to the study of mathematics learning that describe it in terms of students' advance through learning trajectories. According to Clements and Sarama (2014), learning trajectories have three components: (a) a mathematical goal; (b) a developmental path and (c) a set of instructional activities. In Chapter 2 we proposed as a curricular goal of secondary school geometry that geometry instruction can provide means for students to predict what actions will work and explain why some actions may not have worked in students' interactions with representations of figures. That curricular goal takes for granted the possibility to think of the geometric figure as a mathematical concept. We surmise that the geometric figure could play the role of mathematical goal in a learning trajectory. In this second part of this chapter, we closely discuss what development paths can be considered in order to achieve that goal.

We start from a consideration of van Hiele's theory of levels of geometric thinking. While we are aware that there have been substantial improvements as well as critique of the theory (see Battista, 2007), we attend in detail to the original van Hiele theory because it helps provide a background for the research on geometry learning that has led us to attend more specifically to learning trajectories. Van

Hiele's theory of levels in geometric thinking has been foundational in research on geometry learning. This research, in turn, has gathered evidence that questions the notion (fundamental to the van Hiele theory) that students are at any moment at a single stage. Rather, students' level of geometric thinking seems to be sensitive to the specific mathematical goal and contextualized by the activity in which it is observed. The contemporary notion of learning trajectories both crowns the van Hiele-inspired research on learning and pushes the van Hiele theory itself to a place in history. Much as Euclidean geometry maintains a stable position in the secondary curriculum despite the efforts of some reformers of the early and mid-twentieth century to jettison it in favor of more modern approaches to the field, so too does the van Hiele theory continue to play a dominant role in the discourse of students' thinking about geometry, notwithstanding several decades of critique regarding its value and validity as an empirically-grounded theory. Current research on learning in geometry that aims at describing learning as a transition through stages is specific to the geometric objects of study. Some studies have already provided learning trajectories, or alternate approaches to van Hiele levels, for mathematical ideas encountered in primary school (Clements et al., 2004; Clements and Sarama, 2014; Sinclair and Bruce, 2015). Our aim is to provide underpinnings for learning trajectories for ideas encountered in secondary geometry based on the conceptions of geometric figure, mathematical modeling, and informal proofs. An example of learning trajectories for geometric figures is proposed in section 3.5.

Both the van Hiele theory and learning trajectories theory look at learning as a developmental phenomenon by identifying stages in a progression of ways of thinking. Research on geometry learning has also looked at this phenomenon from other perspectives, for example from a discursive perspective in which learning is seen as changing modes of participation in a discourse. We mention these other perspectives here to help the reader mentally organize the field. Psychology looks at learning from the perspective of cognitive mechanism (what internal mechanisms enable students to change from one to another conception) as well as from the perspective of individual development (through which stages and in which order do students progress as they learn about a domain).

Mathematics educators have drawn inspiration from both cognitive and developmental perspectives as they looked at the learning of geometry, and in some cases in more than one way. In terms of accounts of the cognitive mechanisms of learning, both Piagetian scheme theory and information processing traditions have provided means to think about learning. The late work of Piaget, as interpreted by von Glasersfeld (1995), accounts for learning in terms of the assimilation of new stimuli to existing schemes or the adaptation of schemes so that they can assimilate stimuli, and has been quite influential in mathematics education research (e.g., Steffe, 1994). But there is surprisingly little work that describes the learning of geometry in terms of adaptation of schemes, though recent work by Greenstein (2014) is an illustration of this approach in elementary geometry. Also, geometry has played a role in information processing accounts of learning such as Anderson's (1982), who describes learning as the compilation of production rules into larger chunks. This approach to geometry learning has spawned a large literature base on students' learning of geometry using a Geometry Tutor (Anderson, Boyle, Corbett, & Lewis, 1990; Anderson et al., 1995).

In terms of considering learning as development, a classical line of work in geometry learning has been associated with the van Hiele levels of geometric thinking, which describes learning in terms of a transition across stages. As we note above there are aspects of van Hiele's work to be critiqued—notably the notion that individuals will be at a single level of geometric thinking at any one time or the lack of attention to language and other tools in mediating learning. But the van Hiele work has spawned considerable research and it could be considered a precursor of current work on learning trajectories or learning progressions. The next section reviews this work briefly, partly to suggest a basis for investigation into learning trajectories at the secondary level.

3.3.1. Research into van Hiele Levels of Geometric Thinking

3.3.1.1. Van Hiele Levels of Geometric Thinking

The van Hiele levels of geometric thinking (van Hiele, 1957/1984, 1999) originated in the doctoral dissertations of Dina van Hiele-Geldof and Pierre van Hiele, submitted to Utrecht University, Netherlands in the 1950s. As noted above, their work stems from developmental psychology; in particular it builds on Piaget's theory of stages.[5] The 1950s can be seen as "the product of a time when psychology, anthropology and linguistics were redefining themselves and computer science and neuroscience as disciplines were coming into existence" (Miller, 2003, p. 141).

The van Hiele theory provides us with a comprehensive framework to describe students' long-term growth of geometric thinking. This is what Pegg and Tall (2005) called "global frameworks of long-term growth" (p. 469). It suggests that students advance through levels of thinking in geometry with the characteristics of the following properties: "Fixed sequence: A student cannot be van Hiele level n without having gone through level $n - 1$. . . Adjacency: At each level of thought what was intrinsic in the preceding level becomes extrinsic in the current level. . . Distinction: Each level has its own linguistic symbols and its own network of relationships connecting those symbols. . . Separation: Two persons who reason at different levels cannot understand each other" (Usiskin, 1982, p. 5). For many years after being proposed, the van Hiele levels were considered to be a fairly useful model to describe students' behaviors in mainly 2D geometry (Senk, 1989). The model specified the following levels (according to Usiskin, 1982, who uses the 1 through 5 numeration, while some authors number them 0 to 4):

- Level 1: (recognition) The student can learn names of figures and recognizes a shape as a whole. (Squares and rectangles seem to be different.)
- Level 2: (analysis) The student can identify properties of figures. (Rectangles have four right angles.)

- Level 3: (order) The student can logically order figures and relationships, but does not operate within a mathematical system. (Simple deduction can be followed, but proof is not understood.)
- Level 4: (deduction) The student understands the significance of deduction and the roles of postulates, theorems, and proof. (Proofs can be written with understanding.)
- Level 5: (rigor) The student understands the necessity for rigor and is able to make abstract deductions. (Non-Euclidean geometry can be understood.)

(Usiskin, 1982, p. 4)

The van Hiele levels of thinking are not age dependent. Despite their experience of learning geometric shapes at the elementary level, some pupils stay at lower levels of thinking. In the Cognitive Development and Achievement in Secondary School Geometry (CDASSG) project, Usiskin (1982) developed the Van Hiele Geometry Test and conducted a study to test the van Hiele levels of 2,699 students from 13 American high schools. Although Usiskin and his colleagues had difficulties in identifying characteristics of level 5 (rigor), they found that levels 1–4 were generally testable and a reasonable model to describe levels of geometric thinking. Usiskin also concluded that poor performance of proof writing in geometry can be explained by the van Hiele model, as the poor performance of proof writing is strongly associated with lower levels of thinking (p. 89). The CDASSG project results also provided insight into the relationship between van Hiele levels and achievement in writing geometry proofs. Senk (1989) analyzed data from the CDASSG project from another perspective (a sample of 241 students, mainly 16–17-year-olds) and reported that students' ability in writing proof is positively related to achievement in non-proof geometry content tested by the Van Hiele Geometry Test.

3.3.1.2. Studies around van Hiele Levels of Thinking

Since the van Hiele levels of geometric thinking were introduced to the international mathematics education community in the 1970s, many studies have examined the model's capacity to describe students' geometric thinking and measure their geometry learning. A consensus from the studies is that achieving the higher van Hiele levels is difficult for many students.

Other studies also critically examined van Hiele levels. For example, Mayberry (1983) conducted a study with 19 undergraduate pre-service teachers who answered 128 questions during two interview sessions. She concluded that van Hiele levels form a hierarchy, but that students were at different levels for different geometric concepts. Similarly Fuys et al. (1988) conducted a study to examine students' levels of thinking as they enter middle and high school (grades 6 and 9). They found that students' entry level might differ across different topics, though after instruction their levels of thinking might be consistent (p. 180). Burger and Shaughnessy (1986) further attempted to characterize van Hiele levels by using data from clinical interviews. Their participants, children ranging from 6 to 18 years of age, undertook tasks with triangles and quadrilaterals in drawing, identifying and defining, sorting, mystery shape and axioms, theorems, and a proof. As a result of their observation of students solving the problems, the authors could provide more detailed level indicators for van Hiele levels 1–4 (pp. 43–45). For example, at the visual level (level 1 according to our list), Burger and Shaughnessy (1986) observed that students demonstrate "inability to use properties as necessary conditions to determine a shape" (p. 44) but at the descriptive/analytic level (level 2 in our list) they could engage in "descriptions of types of shapes by explicit use of their properties rather than by type names" (p. 44). Gutiérrez et al. (1991) proposed continuous degrees of acquisition levels of van Hiele, and examined the degrees of acquisition of a sample of 50 teachers and grade 8 students who answered their Spatial Geometry Test. Their study shows that this approach allows researchers to describe students who are developing "two consecutive levels of reasoning at the same time, although what usually happens is that the acquisition of the lower level is more complete than the acquisition of the upper level" (p. 250).

Battista (2007) pointed out that there are two important issues in the measurement of van Hiele levels. One is "a lack of distinction between *type of reasoning* and *qualitatively different levels in the development of reasoning*" (p. 853) and the other is assessment (p. 854). For example, Clements et al. (2001) used the van Hiele levels in their study of students' learning of geometry with the Logo programming language and found that it is difficult to classify students in transition from one level to the next (see also Battista, 2011). They suggest that "different types of reasoning—those characterizing different levels—coexist in an individual and develop simultaneously but at different rates and along different paths" (p. 126). For example, van Hiele level 1 is not purely visual, but descriptive and analytic thinking can be observed along with it, and these develop gradually over time (see also Battista, 2007). Lehrer et al. (1998) pointed out that "static descriptions of knowledge (levels) and its attendant emphasis on discontinuities among these levels" (p. 163) are problematic for describing students' geometric thinking (Battista, 2011). A second point to consider is that there is no correlation between age and van Hiele level, hence obstructing any identification of learning with individual development. Fuys et al's (1988) research suggests that instruction can intervene favorably in students' growth in geometry thinking, in particular the use of language such as recalling geometric terminologies or logical language. They also pointed out that students used different language related to their metacognitions about their quality and thinking at various levels (p. 182). For example van Hiele did not assign levels for students' statements such as "Oh, I see a pattern" or "I should prove this, right?" (p. 182). Learning from instruction is not merely the acquisition of geometric terminology but can also imply growth as described by theories of intellectual development: Students can develop ways of behaving that resemble those of mathematicians, or of engaging in more sophisticated discourse with their peers, which is missing in the van Hiele levels of thinking. Lehrer et al. (1998) also criticized the van Hiele model because it does not account for ecological or historical aspects of classroom communities (p. 163), which is one of the influential factors of conceptual development of geometric thinking (Lehrer et al., 1998).

Based on such critique, we share the widely-held view that the field needs to move on from a description of geometric learning in terms of van Hiele levels. In section 3.2 we noted that there are everyday-based, intuitive and visual (spatio-graphical), and scientific and conceptual (theoretical) aspects of geometric figures in students' geometric thinking; the descriptions of levels of thinking suggest that, as students develop their thinking, their control structures pass from being first dominated by visual representations to gradually becoming more theoretical or conceptual. The review of studies around van Hiele levels of thinking also tells us how a theory evolves in mathematics education research: The initial ideas proposed by the van Hieles were tested by Usiskin and colleagues, confirmed by some, modified or refined by other researchers, and through such dialectic attention have become a main theoretical framework of the field to the point that new approaches to describe students' learning of geometry need to be proposed against the background of van Hiele theory. Battista (2007) provided an excellent review of recent developments in van Hiele based research on levels of geometric thinking (pp. 851–853), and concluded that mathematics educators' feeling of validity of van Hiele levels stems from the four phase progression of scientific/mathematical thinking: That is, the progression from perceptualization, conceptualization, and organization to axiomatization "seems like the natural way for individuals to progress from intuitive, everyday reasoning to formal scientific reasoning" (p. 856). Our intention is to explore what aspects it would be necessary to consider in order for us to understand developmental trajectories associated with the four conceptions of geometric figures. In the following section we attend to how the literature on geometric learning has addressed discursive and gestural expressions which were absent in the van Hiele levels of thinking, but which are important in exploring how semiotic registers may develop in the various conceptions of geometric figures.

3.3.2. Extensions to the van Hiele Levels of Geometric Thinking

While the van Hiele levels are hypothesized levels of geometric thinking and were proposed with little attention to tools and other elements of context, researchers have tried to capture other aspects of thinking through observing students' behaviors and listening to their verbal utterances around mathematical concepts. We have seen above that Fuys et al. (1988) pointed out the van Hiele theory's lack of consideration of the use of language related to students' metacognition. One of the extensions of the van Hiele levels has started from the realization that this semiotic mediation, the mediation of thinking by tools and signs (Bartolini-Bussi and Mariotti, 2008), is not merely an expression of meaning but a key constituent of meaning—one could say that geometric thinking is done through activity with tools and signs. Indeed the conceptions of geometric figure described in Chapter 2 include semiotic registers in their characterization. Radford (2014) observes that thinking "is made up of (inner and outer) speech, objectified forms of sensuous imagination, gestures, tactility, and our actual actions with cultural artifacts" (p. 267). In terms of geometric thinking, there are several studies that examine such aspects of thinking, for example the use of gesture and language to interact with diagrams in conjecturing activities (Chen & Herbst, 2013), the role of rhythms in geometric knowing (Bautista & Roth, 2012), and the use of gesture in communicating about motions in 3D space (Morgan & Alshwaikh, 2012). For instance, Bautista and Roth (2012) closely examine how rhythmic body movements emerge when young children who are at levels 1 (Visual) and 2 (Descriptive) are studying 3D shapes, and argue that "it makes little sense to separate cognitive achievements and competencies from the rhythmic performances that we have observed" at these levels of thinking (p. 102). They further identify the roles of rhythms as both resources for and outcomes of geometric activity (p. 103). All these suggest that the changes in levels of geometric thinking might be associated with changes in the kinds of usage of tools and signs.

Some scholars (Sinclair & Moss, 2012; Wang, 2013; Wang & Kinzel, 2014) have attempted a synthesis between the van Hiele theory's levels of thinking and Sfard's (2008) *commognition* framework, in which thinking is examined as realized through communication. Sinclair and Moss (2012) elaborated on the relationships between discursive aspects of students' thinking and the van Hiele levels in order to examine complex aspects of mathematical thinking which the van Hiele model cannot fully capture. Their starting point is the consideration that "*thinking* becomes *a form of communication*, whereas *levels of thinking* become *levels of discourse*" (Sinclair and Moss, 2012, p. 30); their proposal is based on Sfard's (2008) *commognitive* framework in which she characterizes mathematical communications in terms of four aspects: vocabularies or word use, visual mediators, endorsed narratives (description of objects, relations etc.), and routines (repetitive patterns in discourse) (Sfard, 2008, pp. 133–135). From a discursive point of view, Sinclair and Moss (2012) argue that the van Hiele levels of thinking can be put in correspondence with levels of discourse, described as follows (quotes are taken from Sinclair and Moss, 2012, p. 31):

- 1st level of geometric discourse (corresponds to van Hiele Level 1): The discourse of elementary discursive objects ("The word triangle is used as if it referred to one concrete thing. Saying, on two different occasions, that something is a triangle means, for the child, that although at these two occasions she has seen slightly differing pictures, these pictures were images of the same object").
- 2nd level of geometric discourse (a counterpart to van Hiele Level 2): The discourse of concrete discursive objects ("Transformability is still the only criterion for calling two things the same name. An object still cannot have two different (family) names, and in particular, a square cannot be a diamond").
- 3rd level of geometric discourse (a counterpart of van Hiele Level 3): The discourse of abstract objects ("The transformability is no longer the 'official'

> reason for saying that two shapes can be called the same name. Instead, such a decision is grounded in the recognition of a commonality of verbal descriptions of the shapes: Two shapes are considered as deserving one label because they fit the same verbal descriptions—e.g., 'is a polygon and has three sides'—which means that there is discursive procedure (e.g., counting sides) which, when applied to these two shapes will lead to the same result (e.g., number 3)".

Sinclair and Moss (2012) found that "(t)he process of discourse change does not consist in smooth linear transition from one level to another; rather, it involves oscillating between the old and new forms of discourse, resulting in intermediary hybrid forms of geometric communication" (p. 43), and further research is necessary to elaborate causal relationships of shifts of geometric thinking, understanding and communications.

On that count, the question can be raised as to whether we can call them "levels" of discourse. The principles that the van Hieles attributed to their levels—Fixed sequence, Adjacency, Distinction, and Separation—do not seem to hold given Sinclair and Moss (2012) observations. Perhaps this work needs to be read as adding to Clements and Battista's (1992) questioning of the sequential and hierarchical views of geometric thinking offered by van Hiele as too narrowly-constructed to grasp the complexity and malleability of spatial thinking.

Having said this, the investigation of relationships between discursive aspects and levels of thinking, in particular higher levels of thinking, is definitely one of the most interesting future tasks in research into geometric thinking. Wang and Kinzel (2014) took a similar approach, elucidating discourses related to geometric reasoning used by two students, Ivy and Molly, who were considered to be at van Hiele level 3 but whose reasoning, when described from a discursive point of view, seemed more complex to describe than with the tools that van Hiele level 3 would afford. In particular, the notions of *object-level* (characterized as reasoning with concrete figures such as descriptions of static lines, angles and polygons as movable entities under transformations) and *meta-level substantiation routines* (characterized as reasoning with theoretical deduction such as definitions and proofs, which students at van Hiele level 3 are expected to use) are useful (p. 291). Wang and Kinzel (2014) also found that while both Ivy and Molly were both considered to be at van Hiele level 3 by the van Hiele Geometry Test, their *substantiation routines* indicated that their use of reasoning and justifications were different in terms of their participation in discourse. Molly was not engaging *substantiation routines*, and therefore she could not fully understand hierarchical relationships in parallelograms (see Clements et al., 2001).

The study by Sinclair and Moss (2012) examined the role of discourse and learning in the context of dynamic geometry software (DGS, see also 1.4; Goldenberg & Cuoco, 1998). In the next section, before we propose ideas for hypothetical learning trajectories, we examine how the use of the DGS affect the learning of geometry in the contexts of the conceptions of geometric figures, the modeling approach, and informal proofs.

3.4. Enriching Semiotic Registers, Operations, and Control Structures with DGS

3.4.1. The Use of DGS in the Learning of Geometry

The technological functionalities of DGS can provide an additional set of operators (R) for students to use as they solve problems (e.g., dragging). Take the problem of drawing a square; this problem is one of those within the conception of figure as construction in the microspace. Without DGS, operations involved include drawing lines and circles. These are mediated by physical tools such as straightedge and compass, and students can judge whether their drawings are actually squares or not by measuring sides and angles, or reflecting on construction procedures in order to deduce that the sides and angles are all the same. With DGS, students may draw various geometric objects (e.g., points, lines, etc.) and shapes (e.g., polygons, circles, etc.) on a computer screen. DGS likewise enhance the semiotic register (*L*) students use to represent the work (e.g., the traces of moving points can be displayed), and the sets of controls (Σ) used to regulate operations and validate solutions (e.g., if a construction cannot be "messed up" by dragging, it may be correct).

Drawings on the screen can be *dragged* or manipulated dynamically, and students' thinking can be observed through their use of this function. Battista (2008) has noted how the draggable drawings that one can make with DGS require us to complicate the way we see diagrams. To the extent that diagrams are representations of figures, a dynamic diagram can be seen as a set of representations that one can inspect by dragging, thus adding quantitatively to what a paper and pencil diagram would offer to visualization. But to the extent that dynamic diagrams are constructed in the computer using certain tools that make them preserve some relationships, dynamic diagrams are also pseudo-mechanical objects that afford the opportunity to make observations not only *within* each instance obtained through dragging but also *across* those instances. Thus invariants and other properties can be observed against the background of variability created by dragging (see González and Herbst, 2009, for examples of this).

By dragging, students can explore variable and invariant geometric properties of objects and shapes. Arzarello et al. (2002) summarized seven dragging modalities observed in students' work (wandering dragging, bound dragging, guided dragging, dummy locus dragging, line dragging, linked dragging and dragging test). Students exploit these "in order to achieve different aims, such as exploring, conjecturing, validating, justifying" (p. 67); for example, *wandering* and *guided dragging* are used in discovery phases of an exploration, whereas *dragging test* is used to test a conjecture. For instance, students can draw a square using construction as well as measurement tools, but if only measurement tools are used, the drawing can be 'messed up' when it is dragged (Healy et al., 1994). Thus, the *dragging test* can provide further opportunities to refine control structures in order to realize how a square can be drawn attending to theoretical properties, so that it may not be 'messed up' by dragging, etc. This consequently affects students' thinking regarding what operations should be used so that a constructed figure on the screen will pass the *dragging test*. Arzarello et al. (2002) used the following task in order to investigate students' use of these different dragging modalities (p. 69):

> TASK: Let *ABCD* be a quadrilateral. Consider the bisectors of its internal angles and their intersection points *H, K, L, M* of pairwise consecutive bisectors. Drag *ABCD*, considering all its different configurations: what happens to the quadrilateral *HKLM*? What kind of figure does it become?

By undertaking the task above students gradually developed more theoretical reasoning in their control structures, which Arzarello et al. (2002) describe as "a continuous shift from working in Cabri to theoretical reasoning" (p. 69).

Leung (2008) took a step further to investigate the roles of dragging, and explored how to design sequences of tasks and instructions in order to facilitate students' conjecturing and proving in geometry. He observed students seeking invariant properties of geometric figures when they are dragging drawings on the computer screen. He used Marton's *variation theory* (see Runesson, 2006) which advocates that "variation and simultaneity play a fundamental role in discernment" (Leung, 2008, p. 138) as the design principle for problems with DGS. In this framework, four functions (*contrasting, generalizing, separating*, and *fusing*) are identified, and learning tasks can be designed and sequenced in accordance with these functions. For example, consider the problem "Determine all possible configurations of A, B, C and D such that the condition $\angle ABC = 2\angle ADC$ holds" (p. 140). The following sequential activities can be considered (pp. 140–143):

- First, students can *contrast* to see invariant patterns by dragging with the locus tool (this corresponds to *Wandering dragging*)
- Students can then *separate* a specific pattern from what they contrasted in the first stage, leading to conjecture a geometric statement (this corresponds to *guided dragging* and *dummy locus dragging*.

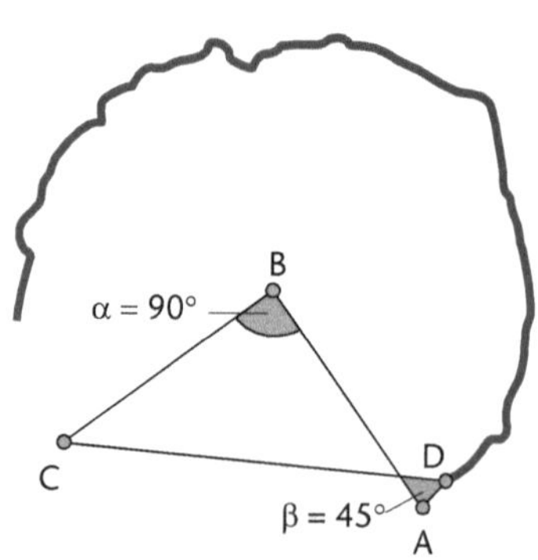

FIGURE 3.3 Seeing invariant patterns

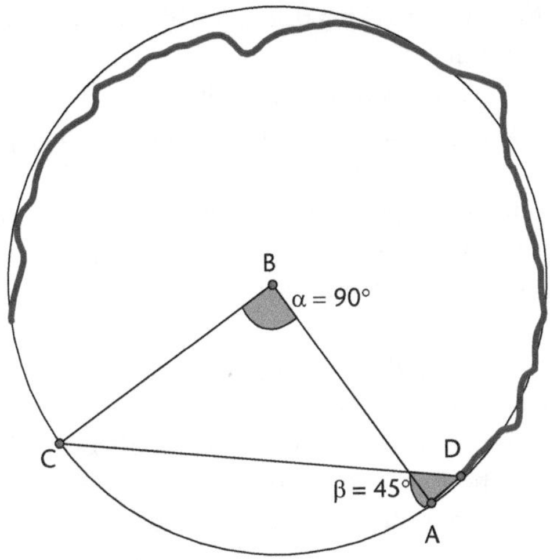

FIGURE 3.4 Separating the pattern

- Students can then *generalize* what they conjectured by constructing more general drawings, i.e., using the *dragging test* or *linked dragging*.
- Students finally *fuse* multiple relationships further by examining critical features of a phenomenon.

The above approach is akin to informal proof in that students start from ideas not formally defined, but using tools from DGS they can reason with those, making conjectures, reducing the problem to simpler problems, to the point of being able to solve it.

3.4.2. Students' Thinking on the Conceptions of Geometric Figure with DGS

As we consider a description of learning as the qualitative improvement of conceptions of geometric figures, an interesting question is: What are the different control structures that participate in different conceptions? How can these control structures describe students' different kinds of thinking when they use DGS? Consider, for example, congruence, one of the key concepts for describing geometric figures. González and Herbst (2009, pp. 155–156) identified the following four control structures present in conceptions of congruence:

- The perceptual conception of congruency (PERC) "relies on visual perception to control the correctness of a solution to the problem of determining if two objects (or more) are congruent".
- The measure-preserving conception of congruency (MeaP) "describes the sphere of practice in which a student establishes that two objects (e.g. segments or angles)

are congruent by way of checking that they have the same measure (as attested by a measurement instrument)".

- The correspondence conception of congruency (CORR) is such that "two objects (segments or angles) are congruent if they are corresponding parts in two triangles that are known to be congruent".
- The transformation conception of congruency (TRANS) "establishes that two objects are congruent if there is a geometric transformation, mapping one to the other, which preserves metric invariants".

The above findings imply that different control structures can be observed in students' work when they tackle problems with DGS, and they could be used to describe transitions from one way of working to another. In what follows we take a few examples from the four conceptions of geometric figures, and show how they can be used to describe students' geometric thinking with DGS in terms of the semiotic registers, operations and control structures.

3.4.2.1. DGS and the Conception of Figure as Description and Manipulation of Small Objects in the Microspace

In our consideration of students' understanding of definitions of geometric figures we have seen different control structures in defining geometric shapes and discussed how students generally prefer partitional over hierarchical definitions, while contemporary school curricula expect them to understand hierarchical definitions. The affordances of DGS described in the previous sections can offer learning opportunities for students to challenge their perceptual or prototypical control structures. For example, Sinclair and Moss (2012) studied young children's (4–5 years old) shifts of levels of discourse when they used DGS to investigate properties of triangles. Surprisingly, with DGSs even young children could engage level 2 and even level 3 discourses as a result of explorations in which they continuously transformed triangles with what Sinclair and Moss (2012) called a *saming-by-morphing* approach:

> It appears that the continuous transformation enabled the children to focus on the invariant property of opposite sides being parallel and, thus, to 'same' a wide range of shapes. In other words, the saming-by-morphing they did by dragging the parallelogram enabled them to identify the invariant property through which the parallelogram is defined.
>
> *(p. 32)*

In a context of the learning of quadrilaterals, Forsythe (2015) showed how the semiotic register of DGS permits students to manipulate two perpendicular bars of adjustable lengths (diagonals of quadrilaterals) in an operation that she calls *Dragging Maintaining Symmetry* (DMS). Forsythe summarized the following observations for describing properties of shapes corresponding to van Hiele levels of thinking (quotes are taken from Forsythe, 2015, p. 204):

- A holistic perception of a shape e.g., commenting "it looks like a kite" (Van Hiele level 1).
- Describing the shape as split into sub-triangles (Van Hiele levels 1 or 2).
- Describing properties of sides and angles, or properties of the bars e.g., "*AB* is equal to *BC* and *AD* is equal to *CD*" (Van Hiele level 2).
- Describing common properties of shapes, e.g., lines of symmetry or congruent sides / angles generated by dragging according to the DMS strategy (Van Hiele level 2).

In addition to the learning under DMS, Forsythe (2015) used dynamic animations under DMS in order to encourage students to pay attention to the continuous nature of shapes as they vary on the computer screen. Her teaching experiments suggested that the study of quadrilaterals with DMS enabled (so as to maintain certain properties of shapes) challenged the students' partitional classification of shapes and led students to recognize a "dragging family" (p. 217) of shapes, helping students shift from the perceptual or prototypical to the hierarchical control structures. Battista (2008) has made similar observations about how DGE such as Shapemakers can help students understand the hierarchical relationships among quadrilaterals.

In a practitioner article, Wong (2013) narrated how he uses dry spaghetti to get students to investigate which combinations of the conditions (e.g., diagonals with equal lengths, diagonals intersecting in the middle, diagonals crossing at a 90 degree angle, etc.) would create which types of quadrilaterals. This sort of task can be replicated with DGS, creating a semiotic register of bars (line segments) that maintain properties such as equal lengths, perpendicularity, bisecting segments, and so on. With the 'diagonals with equal lengths' condition, students can create a variety of quadrilaterals including rectangles (which are parallelograms), but the 'diagonals intersect in the middle' condition creates only parallelograms; and yet these parallelograms can become rhombi, rectangles, or squares as one drags them (*saming-by-morphing*). If students explore both conditions, they might realize that by combining the constraints in both explorations they can create rectangles. In this task, quadrilaterals on the computer screen are mediated by the conditions of diagonals, and the operations they may utilize include choosing the conditions to use, moving the chosen diagonals, and so on. These bars can also provide models of shapes that allow learners to keep in mind alternative possibilities, such as that a parallelogram can sometimes become a square or rectangle, as suggested by the mental model theory (see Johnson-Laird, 1983; Stylianides & Stylianides, 2008). As such, students might move from a perceptual or prototypical control structure to a hierarchical control structure for defining geometric shapes.

3.4.2.2. Modeling Navigational Experiences and Large (Mesospace) Objects in 2D

Earlier we introduced an example of a problem—the amusement park ride (Figure 3.1), taken from a Japanese textbook—within the conception of figure as navigation of the macrospace. The original question—to think why the platform feels level when riders are on it—could be initially modeled by the conception of figure as capture of a large object in the mesospace: One could observe what happens as the ride is being operated and even take a video record of it. Further, as the student looks at the capture on video, the problem may be further translated into the microspace conceptions—first by describing the ride as the transformations of a parallelogram (figure as description and manipulation of small objects in the microspace), later possibly by constructing an operational microspace model using DGS (figure as construction of a small object in the microspace). One might want a parallelogram whose angles can be varied by dragging in order to represent the ride being run. For this latter work, in order to model the problem situation, students not only have to construct parallel lines, but they also have to make sure the diagram they construct would not be 'messed up' by dragging. They can try, test, and modify various methods so that the line that represents where people ride always remains level, and the reasons why; and these movements can be modeled by using the properties of circles with different semiotic registers for this problem (see Figure 3.5 below).

Similarly, Hipparchus' problem, used by Jahnke and Wambach (2013) can also be modeled by diagrams and manipulations with the DGS.

3.4.2.3. Modeling Large Objects in 3D

A useful way to model large 3D objects is to use 2D representations. But, as we have seen earlier in this chapter, students often find it difficult to code and decode

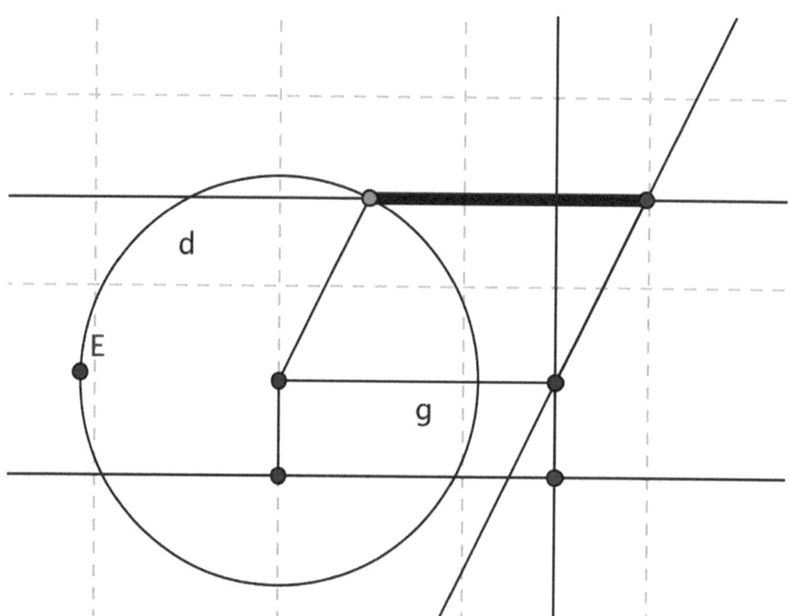

FIGURE 3.5 Modeling a park ride

3D shapes with 2D representations, indicating there are different control structures when students deal with 3D shapes with 2D representations. Meanwhile, Ben-Chaim et al. (1989) demonstrated instructional interventions that can improve students' abilities to communicate spatial information related to a building made up of cubes. While 3D DGSs are still relatively new tools, potential uses and examples have been suggested by several studies (e.g., Christou et al., 2006; Miyazaki, et al., 2012; Oldknow & Tetlow, 2008). Baki et al. (2011) examined the effectiveness of 3D DGS, and showed that the use of the software can help students improve spatial visualization skills, in particular visualizing rotations of 3D shapes with 2D representations, or viewing these representations from various perspectives. They found that the use of physical manipulatives is also effective, but the group that used 3D DGS demonstrated better spatial visualizations. In terms of modeling objects, Oldknow and Tetlow (2008) introduced a range of modeling examples with 3D DGS, including designing a tent, modeling pistons, modeling a reflecting telescope, and so on.

From the point of view of conceptions of geometric figure, the use of DGS with 3D shapes is an interesting area for research on students' thinking and learning. Widder and Gorsky (2013) studied the effectiveness of 3D DGS with two types of problems: "(1) understanding based on given verbal information and logical analytical thinking (a-type items), and (2) understanding based on visualization, that is, the mental manipulation of a given geometric configuration (b-type items)" (p. 97). They identified operations afforded by the 3D DGS tools when students attempt to solve problems such as identifying the shape of faces using measurements, rotations, construction, and painting. They found that students with weak spatial visualization skills tend to use measurement tools but that those with strong visualization skills were more likely to use rotating tools to investigate 3D shapes on the screen, i.e., judging the two representations of 3D shapes through transformations. This indicates that different control structures of 3D shapes in DGS could exist, which might be comparable to the conceptions of congruency identified by González and Herbst (2009) and described above—that is, there may be analogous perceptual, measure-preserving, correspondence, and transformation conceptions for representations of 3D shapes on a computer screen. Ferrara and Mammana (2014) have also shown how some of the features of 3D DGS, such as functionalities that enable a user to see a figure from a different perspective, can provide added controls to support the learners as they make sense of flat diagrams on the screen that are meant to represent three dimensional figures (i.e., the traditional conflict between knowing and seeing, described in Chapter 2).

3.4.2.4. DGS and Geometric Figure as Construction of Small Microspace Objects

DGS provides rich learning opportunities related to the conception of figure as construction of small objects in the microspace. Our rendering of Duval's (2005) notion of dimensional deconstruction emphasizes the cognitive role of construction in supporting transition to non-iconic visualization of geometric objects. The pedagogical value of geometric constructions has been recognized for many decades. For example, Pölya (1957) described geometric constructions as "most suitable for familiarizing the beginner with geometric figures" (p. 3). Students are more likely to have richer understanding of proof if they engage in conjecturing and argumentation rather than in only reading and following prepared proofs (Mariotti, 2000), and geometric constructions could provide opportunities for students to form conjectures and consider why constructions work (Duval, 2005). For example, Pölya (1957), provides the following geometric construction problem:

> Inscribe a square in a given triangle: two vertices of the square should be on the base of the triangle and the two other vertices of the square on the two other sides of the triangle, one on each.
>
> *(p. 23)*

In discussing how to approach solving this sort of problem, Pölya suggests looking for the unknown, the data, and the condition (p. 7). This kind of geometric construction can make geometric learning more active for students. In this case, a student tackling the problem needs to identify *a square* as unknown, *a given triangle* as data, and two vertices of the square should be on the base of the triangle, and the two other vertices of the square on the two other sides of the triangle, one on each as a condition (p. 23).

With the use of DGS, even simple geometric construction tasks can serve such purpose (e.g., Jones, 2000; Mariotti, 2000). Jones (2000) studied how DGS tools mediate students' use of language from imprecise everyday language to more mathematical explanations. The study found that understanding why a diagram constructed with DGS cannot be messed up by dragging (the *dragging test*) is a key criterion students acquire when they progress to more advanced geometric reasoning. Mariotti (2000) also stated that if a diagram constructed with DGS passes the *dragging test*, then that means the construction problem has been solved not only procedurally but also theoretically, as "solving construction problems in the Cabri environment means accepting not only all the graphic facilities of the software, but also accepting a logic system in which its observable phenomena will make sense" (p. 28).

Suppose a problem to construct a rectangle without using the polygon tool is posed within a DGE. Possible operations are, for example, to choose appropriate objects and draw them on screen, to use construction tools to draw perpendicular lines or parallel lines, to use measurement tools to create right angles or sides of a rectangle, to draw circles to take equal distances and so on. The outcomes of such operations are represented through drawings on the computer screen. Reflecting on what we have discussed in Chapter 2 and this chapter, students' predictions and reasoning are likely to be controlled by different control structures. For example, some students construct a rectangle by drawing four lines and perceptually judging that the object can be a rectangle because 'it looks like a rectangle' (the perceptual conception), while others might use measurement tools and use numerical values for sizes of angles or lengths of sides to argue that the figure is a rectangle (the measurement conception).

3.5. Theoretical Underpinnings for Learning Trajectories of Geometric Figures

3.5.1. Hypothetical Learning Trajectories

In sections 3.2 and 3.3 we discussed visual and conceptual aspects of geometric figures in students' geometric thinking; the descriptions of levels of thinking suggest that as students develop their thinking, their control structures pass from being first dominated by visual representations to gradually becoming more theoretical or conceptual. But we have also noted that it is very difficult to capture students' use of semiotic register or system of representation (e.g., those related to metacognitive thinking) in terms of a progressive developmental framework. Rather students' learning of different geometric ideas is not derived from a centralized, homogeneous progress in levels of thinking, but mediated (and differentially so) by the representations used to work with those ideas (representations that include language and tools). In that fundamental sense, the van Hiele levels of thinking are not sufficient to describe learning. The need to overcome problems with the van Hiele theory have supported the development of newer theories that propose that learners go through learning trajectories. As stated in 3.3, the learning trajectories approach is a response to the limitations of the van Hiele approach in the sense that the levels are not homogeneous for given students but might depend on what ideas they are dealing with; this is consistent with the conceptions approach. Learning trajectories still maintain the sense of progression through qualitatively different ways of thinking, though these ways are anchored in specific conceptions and mediated by the elements of those conceptions.

Learning trajectories have three parts: (a) a mathematical goal, (b) a developmental path, and (c) a set of instructional activities (Clements and Sarama, 2014). In this final section of this chapter, we offer ideas for developing learning trajectories for secondary school geometry based on the modeling approach of providing means to predict what actions will work, and explain why some actions may not have worked in students' interactions with representations of figures and the four conceptions of figure. The developmental path that students follow might be described using van Hiele levels of thinking, but more recent research on geometric learning, while bearing some connection to van Hiele levels also departs substantively from it. A main way in which van Hiele levels have been present in this current work is by casting learning as a transition through stages. Scholars like Clements and Battista (1991) who used the van Hiele levels to frame developmental research on students' geometric thinking found evidence to propose modifications to the theory, such as the existence of an earlier syncretic level of thinking and the notion that a person is never completely at one level of thinking but rather that their thinking reveals a mix of levels of thinking. Furthermore, considerations of semiotic mediation of thinking, coming from sociocultural traditions, have been the source of proposals to look at the development of geometric thinking as the development of competency in the use of tools or the development of discourses, including gestures and patterns of communication. In suggesting below learning trajectories for the four conceptions of figure we capitalize on those developments.

3.5.2. Hypothetical Learning Trajectories for Geometric Figure

In elementary schools, children's learning is likely to be based on their practical, hands-on experience such as navigating, drawing, constructing, comparing or describing geometrical objects in macro-, meso- and micro-space. Such activities can be "a matter of making sense in our world as well as making sense of our world" (Thom and McGarvey, 2015, p. 468). The developmental paths in elementary schools with such activities can be based on the trajectory proposed by Clements et al. (2004) in which children developed their understanding of shapes "from a trial-and-error combination of whole shapes through an increasing ability to combine shapes based on their attributes" (p. 167).

We propose an application of the learning trajectories perspective as a theoretical underpinning for students' learning of more sophisticated conceptions of geometric figure in secondary schools. We are not aware of research on adolescent students' learning that has gathered evidence to this effect; rather our proposal is based on modifying what we have learned from van Hiele-based research and from uses of the learning trajectories approach with younger students (Clements and Sarama, 2014). We believe that these learning trajectories are useful for thinking about instruction, particularly to frame the instructional activities proposed in Chapter 5. We also believe these learning trajectories call for research projects that gather evidence to ground them; we hope some readers will be stimulated to do such research in the future.

In all cases, these learning trajectories start from embodied, visual, and holistic thinking that could be associated with lower van Hiele levels and progress to more conceptual, deductive, and analytic thinking, including proving mathematical statements that could be associated with higher van Hiele levels. But by using the four conceptions of figure as starting points of the trajectories, we allow for the possibility that learning may not be homogeneous across them, but rather might progress at different rates or different levels of use of semiotic registers such as languages for different conceptions of figure. Thus, we take an approach similar to that of Clements et al. (2004) who said that their "developmental progression goes beyond existing van Hielian thought in adding the composition and decomposition processes as essential elements of geometric knowledge" (p. 166). We propose four learning trajectories, each of which starts with one of the conceptions of figure introduced in Chapter 2, and each of which progresses in stages to more sophisticated conceptions of figure. Inasmuch as a conception of figure could be unpacked into conceptions of particular figures (e.g., parallelogram, cube), there still exists the possibility that progress along each learning trajectory may not be homogeneous. To avoid the sense that these learning trajectories consist of discrete states of mind and instead promote the notion that learning trajectories consist of conceptions (that is spheres of practice) that demand qualitatively different kinds of performance by students, we describe each in terms of the activities or problems being handled, rather than as levels or stages. They are meant to be organized in order of complexity, and we anticipate that as we (and others) keep working on these trajectories, these conceptions will organize in something like a partial, rather than a total, order.

3.5.2.1. A Learning Trajectory for Figure as Navigation of the Macrospace

Young children exhibit a conception of figure as they navigate the macrospace, actually moving about land and water, within cities, inside buildings, and so on. In secondary school a first activity in a trajectory developed over this conception might involve reading and enacting maps or sets of instructions for a navigation (e.g., the directions provided by a GPS). A second activity in this trajectory might involve representing accomplished navigations of the macrospace, for example by narrating or depicting an accomplished navigation. A third activity in this trajectory might involve anticipating the navigation of a known macrospace, for example by providing verbal instructions or by creating maps. A fourth activity on this trajectory might involve predicting and calculating locations or other properties of navigations based on the reading of maps, narrations, or instructions (e.g., deciding which traversal was shorter).

3.5.2.2. A Learning Trajectory for Figure as Capture of Objects in the Mesospace

Young children exhibit a conception of figure as they deal with large objects (e.g., cars, rocks, other people's bodies) in the mesospace: They seize them by sight, walk around them, climb them, etc. A first activity in a trajectory developed over this conception in secondary school might involve identifying objects or their parts from representations such as drawings or descriptions (e.g., identifying furniture parts from pictorial or verbal assembly instructions). A second activity in a trajectory developed over this conception might involve describing or drawing mesospace objects and their parts, e.g., describing the geometric properties of a ride in an amusement park from an observer's perspective. A third activity might involve constructing scale models (including scale diagrams of projections or perspective diagrams of the object) of mesospace objects. A fourth activity in this conception might involve predicting or calculating the position of mesospace objects in the macrospace and anticipating navigations of the macrospace by mesospace objects different than one's own body (e.g., designing how a large object needs to be moved within a house). A fifth activity might involve designing and creating new parts for mesospace objects or designing how two mesospace objects will interact with each other.

3.5.2.3. A Learning Trajectory for Figure as Description and Manipulation of Objects in the Microspace

Young children exhibit a conception of figure when they request objects by referring to the names of their shapes and when they handle small objects in small spaces (e.g., sorting and matching shapes). A first activity in a trajectory developed over this conception later on involves describing parts of a shape (e.g., sides, edges) or calling up simple manipulations needed to make a more complex manipulation (e.g., flip vertically first, then turn right). A second activity might involve describing parts of a figure or manipulations of a figure more precisely (e.g., through measurement or comparison). A third activity might involve seeing alternative or additional geometric objects in the figure, or related to a figure (including locating objects that were not visible; this may involve transformations). A fourth activity might involve developing a language of description that can systematically handle all communication activities involving existing microspace shapes. A fifth activity might involve producing mathematical statements, imposing or deductively inferring properties or attributes (measures) to the figure on the basis of descriptions or measurements of the figure.

3.5.2.4. A Learning Trajectory for Figure as Construction of Objects in the Microspace

Young children exhibit a conception of figure when they create models or draw diagrams that respond to given names or reproduce small objects holistically. A first activity in a trajectory developed over this conception in secondary school could involve drawing geometric parts of a small object attending to topological characteristics derived from visualization or description. A second activity in this trajectory might involve drawing parts of a small object attending to metric characteristics derived from measurement or comparison. A third activity in this trajectory might involve drawing parts of a figure with theoretical controls derived from properties that might be perceived concomitantly in the figure being constructed. A fourth activity in this trajectory might involve drawing parts of a figure with theoretical controls derived from underlying structures such as the properties of classes of figures that contain the figure being constructed.

3.5.3. What Needs to be Done to Research These Learning Trajectories

It goes without saying that all of the above proposed learning trajectories have been described far too succinctly; furthermore, each stage has been described only in relation to selected aspects of the conception that characterizes it, quite often the kind of problem to be solved. It remains for researchers to flesh out the characterization of each of those stages using the model of conceptions proposed by Balacheff and Gaudin (2010) and described earlier. This work can be done interactively with observations of students at work in tasks that have the chance to bring about the conceptions targeted. The reader will find some suggestions in Chapter 5, but let us here take one example.

One of the learning goals associated with the development of a language of description for microspace objects is to understand and be able to provide definitions of polygons. Earlier in this chapter we have identified three control structures, prototypical, descriptive, and hierarchical; and, associated with the latter, economical constraints on definition (to enable the deduction of other properties from a minimal and most powerful set of stipulated properties). Students' decision-making is likely to be controlled initially by their perception and an important point for designing instructional tasks is to consider how we can challenge such perception, e.g., what operations do students need to do in order to see the shapes more flexibly, and what kind of tasks might perturb their operations so that they could develop their thinking more flexibly? De Villiers (1998) suggests that teaching approaches for developing students' understanding be constructive, and that a progression can be considered from *visual* to *economical*, passing through definitions that might be better than visual but still somewhat uneconomical.

A second task that could be done is to investigate the extent to which a partial order among conceptions can be confirmed empirically. To what extent can one say that a conception only develops after another conception, rather than before, if the distinctions among the conceptions are not made on the basis of different ways to solve the same problem but on the basis of different ways of acting in different problems? Wilkins and Norton (2011) showed how one can make claims about the order in which conceptions are developed by analyzing cross-sectional responses from students to sets of tasks that call for the different conceptions at play. We believe that tasks can be created to assess high school geometry students' performance and that could be used to investigate contingencies of temporal development among the various conceptions that pertain to each of the four learning trajectories. Each of those tasks could assess each of the conceptions hypothesized at each stage in the proposed trajectory. A cross-sectional response to one set of those tasks could produce contingencies of success and failure, and those contingencies could be examined statistically to test the hypotheses that some conceptions are likely to be more basic than others.

3.6. Conclusion

The learning trajectories approach (Clements & Sarama, 2014) has maintained a focus on learning as progressive development and an awareness of the role of the instructional tasks as loci where the learning stage is observed. Yet the learning trajectories approach has relinquished the sense that development is homogeneous within the individual and across mathematical ideas; instead, the learning trajectories approach considers development as specific to particular mathematical goals and attempts to document the conceptions that such learning goes through. The work on learning trajectories has heretofore been anchored in early childhood and lower elementary grades and it would be exciting to see it progress to take root at the secondary level too. The work we propose here, and later in Chapter 5, might offer context for investigations of students' learning trajectories regarding various concepts of secondary mathematics, in particular defining, and producing general statements and proofs. Each of the four conceptions that we identified in Chapter 2 potentially contains ideas that might evolve as students grow in sophistication. The notion that students' development might not be the same across ideas even if it proceeds in steps helps us liberate the notion of increasingly deductive control of what one knows from the notion that one must be able to function in formal axiomatic systems; in particular it presents the option to chart the development of proving capacity in ways that are specific to the conceptions the proof is supposed to help control. Similarly it might be useful to chart adolescents' development of notions of model, definition, and proposition.

We have noted before that the secondary school geometry curriculum could engage students in increasing their means of control over representations of figure in each of the four conceptions of figure discussed in Chapter 2. We noted that one general idea for how to produce such engagement is to involve students in modeling activities, where one conception of figure is used to model problems in the other—we illustrated this using the conception of circle in the microspace to model the problem of the rolling wheel—and in that context we noted there were chances to develop new mathematical ideas such as arc length. We come back to examples like that one in Chapter 5, after dedicating Chapter 4 to an examination of the work of teaching and teachers' knowledge.

Notes

1 A hierarchical definition of parallelogram would define it in such a way that rectangle and rhombus are included among parallelograms, for example by stipulating two pairs of parallel sides but saying nothing about the relative length of consecutive sides.
2 A partitional definition of trapezoid would define it in such a way that parallelograms are not trapezoids, for example stipulating that trapezoids have one and only one pair of parallel sides. Euclid's definitions included partitional ones; for example the definition of the rhombus requires it being equilateral and not right-angled, which excludes the possibility that a square be a rhombus. Both Playfair's (1848) geometry and Brewster's translation of Legendre's (1843) geometry included partitional definitions. A perusal of textbooks from the nineteenth and twentieth centuries suggested to us that hierarchical definitions of "parallelogram", "rectangle" and "rhombus" did not become the norm until the beginning of the twentieth century, and that the partitional definition of "trapezoid" remains the norm in many US textbooks.
3 Note that if rectangles are included in parallelograms, they would have all properties of the latter, including that their diagonals bisect each other.

4 Yet this claim, although widely asserted, is not true unless one makes hierarchical definitions inclusive beyond intuition. Consider the proposition: "The angle bisectors of a rectangle meet in four distinct points which are the vertices of a square." This statement is true if one uses a partitional definition of "rectangle", which excludes squares. But if one uses a hierarchical definition, either the proposition is true only for non-square rectangles, or one needs to consider a single point as a degenerate case of "square" (because the angle bisectors of a square meet at a point).
5 Unlike constructivism, which builds on Piaget's theory of equilibration (or scheme theory; see von Glasersfeld, 1995) to produce a theory of the cognitive mechanisms for learning.

4

TEACHING PRACTICE AND TEACHER KNOWLEDGE IN GEOMETRY INSTRUCTION

4.1. Introduction

The present chapter takes a look at teaching and teacher knowledge in geometry. There is evidence that underlines the general impact of teachers' work on students' learning across learning contents (see e.g., Hattie, 2003). But what is known about the work of teaching specifically in geometry? Studies of the specific work of teaching geometry have been less prominent, though scholars like Rauscher (1993) have claimed that there is a "teacher effect" on students' performance in geometry. This is reasonable, inasmuch as teachers are responsible not only for effecting and mediating the encounter of learners with content, but also with translating the intended curriculum into an enacted curriculum—a daily and weekly act of interpretation that filters the curricular developments of the past century through the lens of teachers' own knowledge, beliefs and values, as well as through the practices that have been customarily used to do that work in school institutions.

But despite the central role that teachers play in defining, reconstructing, and managing the curriculum in their own classrooms, it is only relatively recently (within the past 25 years) that scholars have turned their attention to studying the teaching of geometry at the secondary level. The chapter covers research on the work of teaching in geometry and research on teachers' knowledge of geometry. We start with some framing considerations for the study of geometry teaching, discuss instructional situations in the teaching of geometry, and eventually move to discussing teacher knowledge of and beliefs in geometry.

4.2. Teaching Practice in Geometry

How is the work of teaching geometry shaped by the subject matter being taught? To examine this question we need to frame it under a more general set of considerations—under what assumptions could one say that an account of the work of teaching any subject might be subject specific?

An analogy might be helpful here. Sales—that is, the practice of salespeople—is a human services profession that, like teaching, can be described in general terms: For example, one could describe a sale by using the general language of *goods*, *currency*, *transaction*, etc. But an understanding of what the work of sales entails might be dramatically helped by knowing the kind of sale to be performed: To sell insurance and to sell books, for example, are different cases of sales that, in spite of their accounting commonalities, require different work routines, different ways of handling products and money, and even different ways of acting toward clients. While it is possible that book sales and insurance sales attract different kinds of people to their ranks, it is also quite believable that the nature of the sales object shapes the nature of the sales work to the point that if different salespersons (an insurance salesperson and a book salesperson) do different things these differences are not, of necessity, due to individual differences but likely due to differences in the kind of work they are expected to do, even if some aspects of those jobs are shared. We suggest that it is similar in the case of teaching—that the teaching of geometry implies work of different nature than the teaching of history or the teaching of art. We will not engage in the comparative argument (interested readers can check out Stodolsky, 1988), but rather illustrate the specificity of geometry teaching from a look at cases of the work of teaching geometry. For that, however, a general framing is needed.

4.2.1. The Didactical Contract and Instructional Exchanges

These framing considerations are guided by the interest in identifying resources available for a teacher to organize his or her instruction to support the development of students' means to control intellectually their experiences with shape and space as represented by the four conceptions identified in Chapter 2. In seeking what resources are available, we move beyond the contents of Chapter 3 that dealt with knowledge of students' thinking and of their development. We look now for interactional structures or sociotechnical[1] resources that the teacher can tap into to organize instruction.

The didactical contract is one such resource. Following Brousseau's (1997) theory of didactical situations (TDS), we describe the *didactical contract* as a set of general, mostly tacit norms that stipulate how the teacher and the students are to interact about the content of studies. Under these norms the teacher is expected to do certain things to create opportunities to enable the students' learning of the knowledge at stake, and the students are expected to make use of those opportunities to learn it. Norms that describe in general the mutual expectations of teacher and students vis-à-vis knowledge (its teaching, learning, and use) exist in every classroom. While some norms are rather universal (e.g., in most classrooms it is expected that the teacher will ask students to solve mathematics problems and students can expect to be able to solve those problems using the knowledge they have), others vary depending on what might be described as different contracts. For example, while in some classes students may expect to directly apply what they have most recently learned in the solving of problems, in other classes students may expect they will have to develop new ideas or to combine what they know in original ways.

A particular consequence of the existence of a didactical contract is that teacher and students are involved in transactions or exchanges about knowledge, hereafter *instructional exchanges*. As with other professions of human improvement (e.g., therapy; see Cohen, 2005), the success of the work of the teacher hinges on what the students can do, which teachers may only enable and attest, but not do for students. Teachers are expected to create opportunities for students to work with knowledge; teachers are also expected to recognize in students' behavior whether they have engaged with, even acquired, the knowledge at stake. Furthermore, the need for the teacher to manage those transactions requires the teacher to relate to knowledge in at least two distinct ways: As curricular items, or items of knowledge at stake in a course of studies (e.g., Pythagorean Theorem), and also as task enactments, or informed actions in the context of tasks or other concrete endeavors (e.g., calculating a length using other known lengths; see Herbst and Chazan, 2012, for more on instructional exchanges).

A key role of the teacher in instruction, prescribed by most didactical contracts, is to manage instructional exchanges: To create tasks or work opportunities in which students might act informed by the knowledge at stake, and to interpret those actions in terms of the knowledge at stake (e.g., having or not yet having acquired it). Herbst and Chazan (2012) identify two key ways in which teachers manage instructional exchanges: By negotiating ad hoc norms for students' engagement in a completely novel task (for example through a staged process of devolution of responsibility for various aspects of the task, as described by Brousseau, 1997, p. 33) or by framing the task at hand as a case of other tasks done before, where the norms of engagement may be tacit.

If tasks are completely novel, those exchanges need to be negotiated, and the management of such exchanges can be trying for a teacher. As Herbst (2003) showed, using as an example the case of a teacher managing students' work on a task of ranking triangles by area without using the area formula, when a task is novel a teacher may experience a number of tensions that concern, in particular, how to manage students' relation to the nature of mathematical representations, to the meaning of known procedures in the present problem, and to the goal of the present problem. It would be hard for a teacher and her or his class to sustain interactions over novel tasks for long periods of time; instead one can expect that teachers will tend to modify or normalize novel tasks so that they fit familiar patterns and routines of interaction (Voigt, 1985). But earlier mathematical work provides the teacher with other sociotechnical resources that may save the need to negotiate explicitly what to do.

Herbst and Chazan (2012) use the term *instructional situation* for each one of these frames available in a given course of studies for ushering students into work on a familiar task. Instructional situations are thus work contexts within which students customarily engage with particular kinds of mathematical ideas and provide the teacher with evidence of their knowledge (see also Herbst, 2006). We elaborate on this construct in the next sections as we make it specific for the teaching of geometry.

4.2.2. Instructional Situations in Geometry

Ordinary geometry instruction relies on a number of instructional situations to facilitate students' learning of geometric concepts. By *instructional situations* we mean customary frames for mathematical work in which students' productions can provide the teacher with evidence for the claim that students possess particular items of knowledge at stake (Herbst, 2006): An instructional situation is a local contract that stipulates how a type of task is to be done and what kind of thing the teacher is looking for in its doing.

Instructional situation is a quite general concept, used to describe the mathematical work done in classrooms both in terms of specific tasks that students do and in terms of curricular items targeted with such type of tasks. Indeed the notion of *instructional situation* foregrounds the sense that tasks are, for the teacher in particular, instruments for the instantiation and observation of the knowledge that students are expected to acquire and demonstrate.

4.2.2.1. Instructional Situation as a Fractal Concept

Like the notion of *conception* explained in Chapter 2, *instructional situation* is a fractal concept. It can help describe classes of similar classroom interactions as organized around similar sorts of tasks or pursuing similar types of knowledge at stake. But the notion of instructional situation can also be applied within each of those classes of classroom interactions, allowing us to look deeper into them and detect differences among exchanges that are part of a course of studies.

For example, one instructional situation observed in US high school geometry classrooms is the situation of *exploration* (Herbst et al., 2010), in which a concrete representation of a figure is given and students are to explore its properties by interacting proximally with the representation (Herbst, 2004). In particular, students could be given diagrams like the ones in Figure 4.1, along with a range of measurement tools, and asked to make conjectures about those figures. At stake is students' knowledge of the properties of the figure represented—in the case of the figures represented in Figure 4.1, the knowledge at stake could be the properties of opposite sides and opposite angles in parallelograms. The work students engage in when their work has been framed as exploration often includes proximal interactions with the representation (Herbst, 2004) such as visual inspection, measurement, and reasoning with the particulars of the representation given. In the example, students might measure sides or angles, draw diagonals, extend sides, etc. Their work is also followed by making assertions about objects that

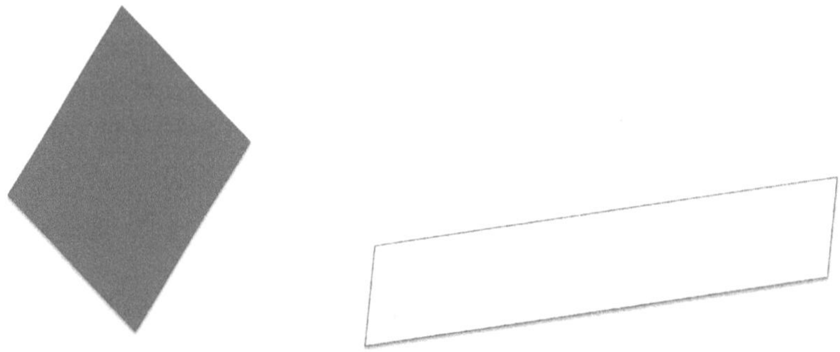

FIGURE 4.1 Representations for an exploration on properties of parallelograms

are part of the figure represented (e.g., "this side measures the same as that side"), which might be translated by them or with the help of the teacher into assertions about concepts that name those objects (viz., "a pair of opposite sides are congruent"). The teacher might take students' observations as evidence that students have (or have not) conjectured the properties of a parallelogram (e.g., that opposite sides are congruent).

Clearly, while the example in Figure 4.1 is specific to the parallelogram, one could anticipate similarly supported explorations of other figures (e.g., rhombi, parallel lines cut by a transversal), which is why *exploration* can index or summon a context for mathematical work without requiring the negotiation of responsibilities that completely novel tasks might require (Herbst, 2003). Thus it makes sense to speak about one situation of exploration, instantiated in each of the explorations of a figure, for a variety of figures. Explorations always frame exchanges between students' proximal interaction with and statements about the particulars of a representation of a figure and the teacher's claim that they have conjectured properties about the concepts involved in the figure.

But exploration could also be seen as a class of situations in that as the specific properties at stake may grow in complexity, the specific work that students have to do may also change. Explorations about the area of figures might require different ways of perceiving than explorations about their congruence; some explorations may call for different tools (e.g., dynamic geometry) with particular technical demands. Because the situation of exploration may unfold into a class of similar situations of exploration when one zooms in, we describe it as a fractal concept. The notion that one could see more similarities when one zooms out and more differences when one zooms in is important not only as a way for an observer to describe the variance in classroom activity (i.e., as featuring a finite set of instructional situations recurring year after year for the teacher, and also every so often within a year of studies), but also as a resource for the teacher to usher the students into work and manage their learning of new ideas and ways of working. The word *exploration* is indexical inasmuch as it may usher students into a task that they know in general how to do; and as they get into it, the particular ways in which that exploration ends up being carried out may expand or differentiate the set of sociotechnical resources with which they engage in exploration.

If one was doing a study of a geometry class through the year, it could be of interest to further unpack *exploration* into various kinds of explorations, and thus further specify the instructional exchanges in such a geometry class, perhaps to the point of describing how specific conceptions are involved in instructional exchanges. And one could note also how the mutual expectations of teacher and students about specific explorations rely on their shared experiences with earlier explorations.

4.2.2.2. Instructional Situations as Frames for Work

The notion of instructional situation is useful for us to explain how instructional interventions designed to promote the evolution of particular conceptions might be anchored in the customary work of a class. The example of situations of exploration introduces the notion of instructional situations as a fractal concept. In order to use it to describe how instructional interventions can be anchored in existing practices we zoom out. An *instructional situation* frames an instructional exchange between canonical forms of work (canonical tasks that students do) and kinds of knowledge at stake (e.g., properties, skills, designated in the curriculum as items to be learned, or goals for students' learning).

The notion of instructional situation attempts to account for the sociotechnical resources (i.e., existing knowledge practices) a teacher can use to fulfill their role in the didactical contract. The didactical contract obliges the teacher to organize students' work and to account for students' accomplished work by recourse to the knowledge to be taught and learnt. A teacher has the job of assigning work to students in which students occupy themselves with particulars (particular diagrams, particular instances of a concept, particular questions); but the teacher also has the job of taking students' engagement in (and completion of) that work not on its own merit but as evidence that students know (or had the opportunity to learn) a certain idea which was involved in their work on the problem.

The completion of particular tasks in school is, more often than not and unlike in many other places, not an end in itself but rather a transactional object. The transactional nature of students' work is particularly obvious for students; as Doyle (1988) noted, tasks have value in the economy of grades and other appraisals students receive—grades, praise, status. But students' work is transactional also for teachers: As we argue above, completed tasks have instructional value inasmuch as they can enable the teacher to make claims about what their students' knowledge is or what opportunity they have had to learn it. These transactions can be fraught with subject-specific tensions for both the student and the teacher, and especially if the tasks are novel (Herbst, 2003).

Instructional situations serve to frame mathematical work so as to mitigate against those tensions. Instructional situations (e.g., exploration) rely on systems of norms that account for the relative predictability of familiar tasks—when students and teacher seem to know what they are supposed to do and what kind of thing they are doing it for. Most mathematical tasks are assigned by appealing to (and against a background of) similar tasks from which students and teachers can implicitly take notice of the role they are to play in the division of cognitive labor. Most of the times that teachers assign a problem, this problem is predicated on a familiar kind of knowledge and resembles other problems that have been posed before; students and teacher thus have some familiarity with what they are supposed to do (see McGowen and Tall, 2010). Various kinds of novelty (at the very least the novelty of specific data given in the problem) can therefore be construed against the background of that familiarity, either tacitly complying with the norms or otherwise negotiating a local departure from the norms.

Instructional situations are situations in a sociological sense, much like one would use the word *situation* to speak of the activities of getting a table at a restaurant, having an annual physical examination at one's primary care physician, or attending a dinner party at someone's house. These examples illustrate also the fractal nature of the concept of situation. For instance, attending a dinner party has some common expectations that apply in general: In the US, guests to a dinner party are expected to bring something such as a dish, wine, or flowers, but depending on the party there may or may not be additional expectations; attendants to a dinner party for a small number of people who will sit at a table are expected to meet, greet, and talk to every other guest, but if the party is large and guests don't sit at a common table there is no such expectation. Thus the dinner party situation might further subdivide into different kinds of dinner parties as soon as one zooms in. Yet dinner parties are not so different from each other as to be just as different as they are from other ways of having a meal (e.g., family dinner at home, dinner at a restaurant). Furthermore, people use their experience in some situations (e.g., dinner parties at one's parents' house) to manage their way around related situations (e.g., dinner parties at one's employer's house) in such a way that new socializations build on earlier ones. To say that an instructional situation frames an exchange means that the particular work to be done and exchanged is seen as an instance of other kinds of work, familiar to participants and gathered under the name of the situation.

Sociologists have described classroom interactions in terms of generic situations such as *recitation* (Mehan, 1979) and used the expressions *activity structure* or *activity type* to describe generic divisions of labor in these interactions (Lemke, 1990). Our choice to employ the expression *instructional situation* instead aims at underscoring the transactional, instrumental nature of those situations in the procurement of the teaching and learning of particular subject matter, which in our case is geometry. Similar to Brousseau's (1997) notion of didactical situation, an instructional situation frames or contains a scope of work for the student that has been chosen on behalf of an item of knowledge at stake: The instructional situation contains expectations for how the teacher will set up and handle that scope of work as well as how the student will accomplish that work.

Thus, instructional situations frame familiar tasks. In the next chapter we will see that they are useful as sociotechnical resources for instructional interventions.

The existence of instructional situations creates conditions of possibility for novel tasks to be deployed in ordinary classrooms, anchored in the practices of ordinary classrooms. A novel task might initially appear to be like the canonical tasks of an existing instructional situation, but one in which some of the norms or expectations of the situation may not be fulfilled. As the task is pursued, participants may realize the need for negotiation of how the didactical contract applies. In this chapter, however, we use the notion of instructional situation to describe the practices of teaching geometry in subject-specific terms.

4.2.2.3. Different Instructional Situations in Geometry

As we have noted, instructional situations enable exchanges between work and the knowledge at stake. We think of this knowledge at stake not solely as topics (e.g., a particular definition or theorem) but as types of knowledge (e.g., definitions, theorems) and propose that instructional situations exist in geometry classrooms so as to routinely facilitate the exchanges between types of knowledge and types of student work. Some of these situations have been observed in American geometry classrooms: They include *calculating a measure*, *constructing (a diagram)*, *exploring a figure*, and *doing proofs* (see Herbst et al., 2010). We expect that geometry lessons in other countries may include those or other instructional situations and that readers will take the examples of these four instructional situations to study other instructional situations that might exist in other countries.

In general, to describe an instructional situation we ought to make a number of hypotheses that describe what knowledge is at stake and what work needs to be done. The knowledge at stake is whatever curricular claim the teacher could make on the basis of the work done—whatever item of knowledge or skill the teacher could say has been addressed by way of the work done. The work done is the sequence of instructionally-relevant interactions between teacher, students, and content that, in the aggregate, constitutes evidence for the teacher to lay claim on the knowledge being at stake. Obviously, for a teacher to consider that a given object of knowledge is known, they may need to observe several bouts of the same instructional situation; this recurrence, a byproduct of cultural understandings of what it means to teach, to learn, and to know, is partly why the concept of instructional situation is useful.

The American geometry curriculum, like many curricula around the world, delivers knowledge of geometric figures and their properties. Geometry instruction uses a number of instructional situations to enable that communication of knowledge. The situation of exploration of a figure is an important one, but there are several others. Take for example the situation that Hsu (2010) calls "geometric calculation in algebra" (GCA), which is one of those that appear as one zooms into the situation of "calculating a measure" identified by Herbst et al. (2010). In late elementary school students start calculating measures as a way of finding things out about a figure in ways that are different than by exploration: Students are given diagrams with some numerical information about the size of their sides or angles and asked to find out other information, through calculation which sometimes is facilitated by a formula and other times by known properties. Hsu (2010; see also Hsu and Silver, 2014) documents in detail how seventh grade Taiwanese students engage in "geometric calculations in number (GCN)" (see also Küchemann and Hoyles, 2002).

In the US high school geometry course students study geometric properties of figures in the context of *geometric calculations in algebra* (GCA; Boileau and Herbst, 2015). Students are assigned problems where a geometric diagram is provided along with algebraic expressions that represent the lengths of sides or the measures of angles in the figure. Students demonstrate knowledge of the properties of the figure by setting up equations that use the algebraic expressions; they then solve the equations to find the actual dimensions of the figure. What is at stake in this situation is students' knowledge of the properties of the figure, though the work they do is to set up and solve equations. This illustrates the transactional nature of student work and how the exchange between work and knowledge is often of seemingly unequal elements: When asked to find the lengths of the diagonals of a rectangle such as the one given in Figure 4.2 the student may be writing the equation $2x + 3 = 3x - 2$, but the teacher might see in that writing evidence of the students' knowledge that diagonals in a rectangle are congruent.

In order for such exchanges to happen there are tacit norms that regulate what work needs to be done, how, when, and by whom. For example, it is usually the teacher or the textbook that provide the algebraic expressions to be used in setting up the equations, and those expressions usually are such that, when the student solves the equation, the solution will describe an actual rather than an impossible geometric figure.[2] Students interact with the diagram by attributing to it the properties from the givens of the problem and by reading off the diagram the information about its given dimensions. Students use what they know about the given figure to pose an equation that relates some of the algebraic expressions given and they eventually solve the equation using methods learned in algebra, producing values for the variable as well as for the unknown dimensions of the figure.

The way students interact with the diagram in a situation of calculating a measure contrasts markedly with how students interact with diagrams in a situation of exploration. In situations of exploration, what is at stake is also knowledge of

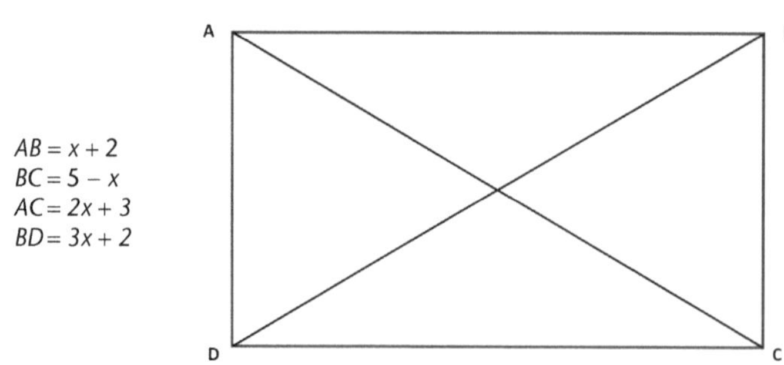

FIGURE 4.2 Calculate the lengths of the diagonals of the rectangle *ABCD*

properties of figures, though these are at stake in a different sense than in situations of calculation: While in situations of calculation what is at stake is whether students can recognize the usefulness of a known property in answering a question about a specific figure, in situations of exploration what is at stake is whether students can abstract the statement of a general property based on the study of specific figures. In situations of exploration, students might also receive one or more representations of the figure (e.g., diagrams) from the teacher but they don't limit themselves to reading off the information given with the diagram; instead they also interact proximally with the diagram, for example measuring the length of segments or the amplitude of angles. Their job is to translate the rather specific information found (e.g., that two strokes have specific lengths) into conjectures of properties about the represented figure (viz., that the lengths of the two segments represented by the strokes stand in a relationship such as shorter, equal, or longer), and perhaps also into conjectures of properties of all figures in a given class (viz., that the lengths of corresponding segments in comparable figures stand in the same relationship).

Herbst et al. (2010; see also Herbst, 2004) describe two other instructional situations where interactions with diagrams are different. In situations of *doing proofs*, like in situations of calculation of a measure, students interact distally with the diagram, neither measuring nor adding to the diagram, though they do rely on some visual inspection of the diagram to propose properties of the figure that might be stated as part of the proof if they could also be justified on account of what is known about the figure. All three of these situations—exploration, calculation, doing proofs—appeal to one of the conceptions of figure described in Chapter 2: Figure as description of an object in the microspace.

A fourth instructional situation, *construction*, appeals to the conception we have labeled *construction of an object in the microspace*. In situations of *construction*, as in situations of exploration, students interact proximally with the diagram, adding objects to it, but the reading off of properties (e.g., from the traces of auxiliary lines or from measurement) serves to control that the objects added are correct (hence the diagram represents the figure that it is meant to represent) rather than to discover properties of the figure. The kind of tools involved in the construction (whether only straightedge and compass, or also ruler and protractor, or DGS instead) allows for the situation of construction to subdivide into several situations of construction, along with the fractal interpretation of the notion of instructional situation.

All of these instructional situations (calculating a measure, construction, doing proofs, and exploration) involve a task that students are given responsibility for doing. As Herbst and Chazan (2012) show, it is possible to reconcile the a-didactical situations of Brousseau (1997) as particularly interesting cases of the tasks that students do to enable instructional exchanges. We note that instructional situations describe recurrent ways in which instructional practice is organized, and that departures from instructional situations, for example in the form of novel tasks (or a-didactical situations), identify singular moments where special learning exchanges may happen.

4.2.2.4. Dynamic Geometry Software in the Teaching of Geometry in Secondary School

We dedicate a special section here to discuss the role of DGS in geometry instruction. The role that DGS has played in revitalizing the curriculum and enabling students to think and learn differently about geometry has been discussed in earlier chapters. In this section we are concerned with how these tools have become part of the didactical contract for secondary geometry. On the one hand the existence of instructional situations of construction and exploration might suggest that DGS has had natural places of deployment. As noted in Chapter 3, teachers have engaged students in exploring dynamic sketches provided to them, expecting that from such explorations students would develop conjectures about figures (Battista, 2008; Laborde, 2007). Also, teachers have used DGS to engage students in constructing diagrams and have used dynamic affordances such as dragging to give students feedback about the necessary properties of the figure they constructed (Hoyles and Noss, 2003). But these activities have for the most part been observed in intervention settings, while descriptive studies have suggested that a more common use of DGS is to enhance more customary aspects of instruction such as teachers' presentation of new material or demonstration of solutions to problems (Ruthven et al., 2008). Sinclair and Yurita (2008) studied the changing aspects of discourse of a teacher who used DGS. They reported that after the introduction of the DGS in a grade 10 classroom, classroom discourse about quadrilaterals and their inter-relationships changed into more 'dynamic' discourses. For example, they noted that "the word 'square' might as well be called 'all-squares-we-could-make' (thus casting the square quasi-empirically rather than analytically)" (p. 147). These observations foreshadow the need to describe more generally how it is that DGS has been incorporated into the didactical contracts that enable teachers and students to work on geometry together.

The use of DGS was first studied from the point of view of its impact on the practice of established teachers who got acquainted with the new technology and made its use part of their teaching praxeologies. The decision in favor of the implementation of dynamic geometry software may cause subtle transition problems. Hannafin et al. (2001) reported on a two week teaching experiment with a class of seventh graders in which the teacher had decided to give more freedom to learners during their sessions with DGS; however, in the end, the teacher had difficulty relinquishing control in the way planned.

Qualitative studies indicate that beliefs might play an important role in the way teachers use dynamic geometry software in their classrooms (see more about beliefs later in the chapter). Shafer (2004) closely followed two teachers from the acquisition of basic technological content knowledge on dynamic geometry software to the actual use of it in their classrooms. Shafer (2004) investigated links between teachers' beliefs and the implementation of the DGS in the classroom. The first teacher, Alice, had an instrumentalist view of mathematics (as described by Thompson, 1992) and fit the category of an "accidental tourist" in the sense of Saye (1998). Teachers with these characteristics have the goal of efficiency and control and stick to those topics with which they are very familiar; technology is typically used to save time and to improve performance. In fact, several observations indicated that Alice avoided using DGS in areas with which she was less familiar and tried to reinforce her way of teaching with the help of the software. In contrast, June, the second teacher Shafer investigated, had a process view of mathematics (Thompson, 1992) and showed the attitude of a "voyageur" (Saye, 1998). She used DGS at an early stage, even with the offered guidance of the tutor, and applied it as a "tool for discovery" (Shafer, 2004, p. 164). The study also revealed a difference in the van Hiele level both teachers addressed during the observed lessons: Even though Alice was tested to be at van Hiele level 4 herself, she remained merely on the procedural level with all students; June, on the other hand, moved between different levels.

How widespread is the use of dynamic geometry software actually? Hendriks et al. (2008) systematically surveyed EU educational systems and suggested that the "availability of DGS in schools does not automatically imply their use" (p. 3). In none of the 27 countries was the proportion of teachers using DGS regularly estimated higher than 30 percent. Müller (2006), for instance, found that among 250 mathematics teachers in Austria about 40 percent had used DGS in their classrooms once, but only one sixth of them continued to use it regularly. "Why don't all maths teachers use dynamic geometry software in their classrooms?" asked Stols and Kriek (2011). Their study of 22 teachers indicated that the "perceived usefulness" of dynamic geometry software was the most important indicator for its use.

In Australia, Handal et al. (2013) found indications of growing technological content knowledge on dynamic geometry software. Teachers gave a moderately high self-assessment concerning the statement "I am able to use dynamic geometry software (e.g., GeoGebra, Geometer's Sketchpad, Autograph, Cabri)" (pp. 28–29); the variance for this question was among the highest in the survey.[3] Concerning the technological pedagogical knowledge and the technological pedagogical content knowledge, dynamic geometry software was not mentioned explicitly. Interestingly, the technological pedagogical content knowledge scale item "I am able to use technology to assist students to develop their maths problem solving skills" (p. 29) was ranked with a high qualitative descriptor.[4]

In Chapter 1 we noted what a historical event the development of dynamic geometry software has been for mathematics education; Chapter 2's discussion of the role of diagrams has provided rationale for its importance in the curriculum; and from Chapter 3, it is apparent that dynamic geometry software has provided a great resource for learning, which has been observed in designed settings. The discussion in this chapter has noted that dynamic geometry software has been usable in the context of situations of exploration and construction, albeit somewhat challenging the work of teachers. We take up this theme again in Chapter 5, when we suggest some design research projects that involve DGS in the modeling approach espoused in this book. But before that, the last part of this section steps back to look at the practice of teaching geometry as a whole, while the sections that follow dwell with issues of teachers' personal resources including knowledge and beliefs.

4.2.2.5. Is the Work of Teaching Geometry Subject Specific?

One could say that, to the extent that one teaches geometry, that work is as a matter of course subject specific. Yet, we would like to think beyond the obvious fact that the object of studies is a domain of mathematics and against the background that much pedagogical theory over the decades has spoken of the work of the teacher as capable of description by eliding the subject matter (e.g., Hattie, 2009, pp. 161–236). The prior discussion about didactical contract and instructional situations provides a framework for arguing that the work of teaching geometry is subject specific, beyond the obvious fact that the object of studies is a domain of mathematics.

Instructional situations, in particular, call for the geometry teacher to do some things to facilitate learning that other mathematics teachers or teachers of other subjects do not need to do. This observation can be tricky to make: Inasmuch as abstraction may allow us to see the same work being done in two very different manifestations, it would be possible to cast the work of teaching in such abstract terms that the differences across the teaching of different subjects might be elided. But the notion that the teaching of mathematics involves specific knowledge that aids teachers in doing their work in specific instructional situations, knowledge that is either possessed by individual teachers or recognized by teachers as being demanded by specific work, helps us discourage the use of such abstractions to describe the work of teaching. We elaborate this paragraph in this section.

Consider two instructional situations—exploring a figure and calculating a measure. In both of them the teacher is called on to create a representation of a figure for students to work with. In the case of exploring a figure, inasmuch as students will interact proximally with the representation and use those interactions to make assertions that instantiate target properties, we conjecture that in order to enable students' mathematical work the teacher would have to take great care in the creation of an accurate geometric representation, with preference to those that would not contain great risk of reading errors. This might mean drawing the representation with precise tools and thin strokes, as well as doing as much as possible to have measurements that are whole numbers or that involve common fractions (e.g., it is more likely that two congruent sides in a parallelogram will be 4.5 each rather than 4.55 each if students are expected to conjecture their congruence from measuring them). These actions on the part of the teacher might be interpreted as extreme attention to detail on the part of the individual teacher but they might also be read as what needs to be done to enable students to use their proximal interactions with the figure to read target instances of the property at stake: If the diagram is accurate, the student would be able to consider the target property as a possibility; if the diagram is very accurate, the student would be able to confirm empirically their perception when they interact proximally, and would be encouraged to state the property even if still speaking of particulars (e.g., "these sides are congruent"). We contend that such attention to detail in creating a diagram for an exploration is an example of how the teaching of geometry is subject specific: The mathematical work students need to do with the diagram makes demands on what the teacher needs to do to set up such work.

In the case of calculating a measure, the teacher also needs to provide a representation, but this one is different and calls for different work on the part of the teacher. The teacher must create a diagram, but since the diagram will not be handled proximally, its drawing does not need to be very accurate. The drawing needs to be accurate enough for the student to perceive it as plausible to read some properties off the diagram, but such readings are not to be confirmed by measurement. So, for example, if a teacher gives students a figure like Figure 4.2., it will be important that segments $\overline{AC}$ and $\overline{BD}$ appear congruent though they do not need to be exactly the same length. It will be important, however, that the teacher also worries about other representations provided—the algebraic expressions that represent the said congruent segments need to be such that when equations are posed and solved, they will provide consistent measures. What is meant by *consistent measures* is heavily subject specific. It means in particular two things: One, that variables need to be consistently evaluated—that is, if x will take a value for one expression, it will take the same value in all expressions that use x to represent dimensions in the same figure. It means also that findings about the measures of the figure obtained after equation solving, need to cohere with other geometric properties that are true about the figure they avowedly represent, e.g., a length cannot be negative or the legs of a right triangle cannot be longer than its hypotenuse.

Clearly, one could say that those cases of the work of teaching are just examples of the teacher creating the givens of a problem for students to be able to work on the problem, and even more abstractly, that those cases are just examples of the teacher creating the resources for an assignment so that students can do the assignment. But students could engage in the work even if the teacher did not create those representations in the manner described: A teacher could draw a diagram sloppily and get students to explore it; a teacher could attribute inconsistent algebraic expressions to the sides of a figure and get students to calculate the measures of its sides. That is, the actions the teacher does could still be described generically as creating the givens of a problem but be done in ways that breach the norm of the instructional situations they support. These breaches would impact the quality of the mathematical work students engage in. This observation suggests that the actions a teacher does need to be described in a subject-specific way, where the categories of subject specificity are identified by the norms of the instructional situation. To be clear, if we adopted a generic description and said that these are just cases of the teacher creating resources for a student assignment, we would need to accommodate within that description both (1) the case of a teacher that does so complying with the norms of the situation and (2) the case of a teacher who does so but breaching a norm (e.g., providing a diagram for an exploration but the diagram is inaccurate). To the extent that practitioners notice (or repair) breaches of norms like these (see Herbst et al., 2013, for an example), we can say that at least for US teachers, the grounds for the distinction we have made are not just different examples of the same generic task of teaching, but actual information, in the sense of Bateson (1972), "a difference that makes a difference" (p. 315). It should also be noted that the examples and arguments offered are from US-based research because that is where these ideas have been grounded so far; yet the notions of instructional situation and norm can be implemented to study the teaching of geometry in other countries, where it is quite possible one might find other differences that make a difference.

The prior paragraphs illustrate the elements of an argument for the claim that the work of teaching geometry is subject-specific. Obviously, the criteria used to detect differences, whether these are summative measures of achievement and success or analyzes of the qualities of the mathematical work, matter in deciding whether these are differences that make a difference. And obviously also, some of the subject-specific differences that the notion of instructional situation permits us to detect are nested in general approaches to teaching (e.g., problem based instruction, direct instruction) that contribute by themselves to making or not making a difference (see Hattie, 2009). But having said that, to the extent that the work of teaching involves transacting students' work on tasks for claims on what they know or don't know, some broad tasks of teaching emerge (e.g., creating work assignments, interpreting the students' work) that are intrinsically connected to the subject-specific work that students do. The way a teacher carries out these tasks of teaching could be idiosyncratic (e.g., a teacher might always be sloppy in the assignments he or she provides), but we would not expect such idiosyncratic behavior to apply to the majority of professionals. Instead we would expect that participants' recognition of the norms of the instructional situation that frames their work, and their knowledge of the mathematics needed to enact such instructional situations, would account for part of the variance in the ways teachers enact these tasks of teaching. As can be noted, teachers' knowledge of the mathematics they teach, geometry in this case, can be crucial to explain what teachers do. Our next sections elaborate on this.

4.3. Teacher Knowledge of Geometry

This section is devoted to what is and is not known to researchers about what teachers know, how they learn it, how they think, and how those factors shape and influence their practice in the context of secondary geometry. The notion that teachers' knowledge of their subject matter is an important variable in the study of teaching is not new (see Shulman, 1986a, 1986b), and it has been elaborated in general for mathematics by many researchers (see Ball et al., 2001; Depaepe et al., 2013). But scholars have also argued for the need to consider specifically the knowledge of geometry for teaching. Hence, Chinnappan and Lawson (2005) showed a way to use concept maps to examine how individual teachers hold their knowledge of geometry. Also, Kuzniak and Rauscher (2011) have shown that the teachers' own conceptions of geometry and its teaching and learning make a difference to their reactions on students' learning difficulties.

Our goal is to answer the questions:

1. What and how have we learned about teachers' knowledge in the context of secondary geometry?
   and
2. What future directions can research in this field be expected to take in the near future?

4.3.1. Teacher Knowledge of Geometry from Professional Preparation Programs

Perhaps the simplest approach to an investigation of teachers' knowledge of geometry is through a study of the courses that teachers take as part of their preservice education and training, and indeed there is some merit to this approach. The curriculum of teacher education may be regarded as a kind of institutionalized answer to the question of what future teachers of geometry should know about the subject they expect to teach, though the institutional answers may vary considerably (Grover and Connor, 2000). From this point of view, a description of preservice teachers' course-taking in geometry may be regarded as one among many possible articulations of what we shall refer to below as mathematical knowledge for teaching geometry, or MKTG. We therefore begin this section with a brief review of what is known about the curricula of preservice teacher education in the specific context of geometry.

In recent years it has become widely recognized that course-taking is a poor proxy for measurements of teacher knowledge. That is to say: It is not nearly enough to know whether teachers have studied this or that course in college, or how many such courses they took, or what textbooks they used in those courses. But as noted above, we may regard the mathematics curriculum of teacher education as providing a kind of institutionalized answer to the question of what teachers *should* know about the subject matter they intend to teach. After all, if a particular aspect of that knowledge seems to be found rather consistently in teacher education programs, that presence would seem to indicate an institutional consensus about the importance of that knowledge. It is a bit more problematic to make the converse argument: If scholars, practitioners, and policymakers believe that a certain type of knowledge is important, might we reasonably expect to find that knowledge represented in the curriculum of preservice teacher education? Different perspectives exist, with some proposing that teachers study the mathematics they will teach and others proposing that teachers study more advanced mathematics that would enable them to complete their preparation to teach on their own (see Ferrini-Mundy and Findell, 2001; Meserve, 1959; Wu, 2011).

For those reasons there is value in attending closely to the geometry content of secondary teacher preparation programs. Such attention is not new; as far back as 1941, one critic noted:

> At present many teachers of high school geometry have never studied any geometry beyond the course they are teaching, except analytic geometry. The colleges place much more emphasis on algebra than on geometry, although there is no justification for offering poorer preparation in one branch of mathematics than in the other. It is a shame for higher institutions that prepare secondary school teachers to send them back to the high schools knowing no more about the subject [geometry] they are to teach than when they left high school. In fact, they know less, for they have had several years to forget what they once knew.
>
> *(O'Toole, 1941, pp. 319–320)*

The distance between the geometry that teachers teach and the geometry they learn in college mathematics courses (an issue already noted by Klein, 1939) continues to be a matter of discussion. Wu (2011) noted,

> So long as university mathematics courses do not address issues arising from school mathematics, teachers will not be sufficiently well informed to reject such mathematical illiteracy, and publishers will continue to get away with the promotion of this kind of illiteracy. We must create a university mathematics curriculum for prospective teachers to help them look back at such school concerns as the meaning of congruence and similarity and why these concepts are important in mathematics. By contrast, preservice teachers are given at least some access to such topics as the curvature of curves, Gaussian curvature of surfaces, finite geometry, projective geometry, non-Euclidean geometries, and the foundations of geometry. They are not, however, taught plane Euclidean geometry. This last is exactly what teachers need because it is usually taught poorly in schools. They desperately need solid information about school geometry in order to better teach their own geometry classes.
>
> *(p. 378)*

The question of how teacher education institutions provide an answer to the question of what geometry should be taught to preservice teachers seems to be open also internationally. In their study of mathematical preparation across six countries, Schmidt, Houang et al. (2008) noted that the preservice mathematical preparation of teachers in geometry might include topics on differential geometry, non-Euclidean geometry, and topology, while courses in pedagogy would cover geometry as a topic of study in school.

Grover and Connor (2000) studied the characteristics of teacher preparation courses in the United States with respect to their geometry content. They note that undergraduate college courses play a more crucial role for those entering the teaching profession in the United States than for those in many other countries because "U.S. teachers typically receive little or no official mentoring after they are hired as full-time teachers" (p. 47). In a spirit reminiscent of O'Toole's (1941) earlier critique, they note that "In the United States, [prospective secondary mathematics teachers] studied geometry only once in secondary school, and they will encounter geometric concepts only once more in college before they are certified to teach" (p. 48), and that it therefore constitutes a very limited window of opportunity to help "prepare prospective teachers for the complex and multi-faceted task of teaching" (p. 49).

Grover and Connor (2000) conducted a survey of faculty members who had recently taught a geometry course for preservice secondary teachers at 108 mathematics departments in US universities, supplemented by an examination of the syllabi of 30 such courses. Grover and Connor (2000) found that the courses in their sample varied widely in terms of their choice of textbook, content, and emphasis. In particular they found that roughly half of the courses focused mainly on Euclidean geometry, while half focused equally on Euclidean and non-Euclidean geometry; that roughly half of the courses employed synthetic approaches, while analytic and transformational approaches were far more limited in use; and that instructors of those courses tended to report a higher degree of overlap with the content of a high school geometry course than the researchers themselves found through their own coding. They also studied the extent to which those courses were aligned with the content standards, and the recommendations for good mathematics teaching for preservice courses, in the National Council of Teachers of Mathematics *Professional Standards for Teaching Mathematics* (NCTM, 1991). Overall they concluded that "wide diversity exists in the courses with respect to the content addressed, the way in which mathematical ideas are developed, the instructional approaches used, and the assessment practices employed" (Grover and Connor, 2000, p. 61). They regarded about 100 geometry courses in the United States as non-traditional. Tatto and Senk's (2011) international study surveyed preservice teachers in terms of their opportunities to learn specific domains of mathematics, including geometry, and found wide opportunity to learn though uneven performance in geometry. Whiteley (1999) argues that as Euclidean geometry declined as a subject of mathematical research, its teaching at the university level was more frequently done by logicians and philosophers (and we could add mathematics educators) than by practicing mathematicians; this arguably might have impacted the extent to which the course of geometry for teachers engaged preservice teachers in the practices of geometric research (Nathalie Sinclair, personal communication, April 2016).

Although it was limited to the United States, Grover and Connor's (2000) study is important in that it specifically targets future secondary teachers' knowledge of geometry. It is not, however, the only study to have directed attention to the practices and purposes of such a course. For example, Blanton (2002) describes her use of an undergraduate geometry course that she designed with the intent of preserving its mathematical integrity while simultaneously challenging preservice teachers' expectations about discourse in a mathematics classroom. Blanton (2002) looks to Lortie's (1975) recognition that teachers' own experiences as students constitute an "apprenticeship of observation" that powerfully shapes their practices as teachers, and to Grossman's (1990) observation that those practices are likely to reflect their most recent experiences in the undergraduate classroom, in her attempt to investigate how an undergraduate mathematics course could "be used to heighten students' awareness of and ability to engage in and cultivate a discourse that promotes a conceptual understanding of mathematics" (p. 119).

For the purposes of this chapter, the significance of Blanton's study lies in its identification of a type of geometry knowledge that is distinct from the usual conception of content. Blanton observes that "students seemed to be learning how to think and argue mathematically *by the sheer act of doing it*" (p. 139, emphasis in original), and argues that in such courses students can learn not only the definitions, postulates, theorems, and results of geometry but also the discursive practices that can support more robust and student-centered ways of teaching mathematics.

Courses in geometry for teachers, such as the ones found in US teacher preparation programs, are not the only setting in which preservice teacher education calls attention to teachers' knowledge of geometry. Tsamir (2007) describes a set of interactions in a course entitled "Psychological Aspects of Mathematics Education," required of all preservice secondary mathematics teachers at Tel-Aviv University in Israel. Tsamir (2007) describes a class in which prospective teachers learn about the informal (and normally tacit) *intuitive rules* (Stavy and Tirosh, 2000; see also Tirosh et al., 2001) that often typify students' erroneous thinking. The narrative continues to focus on a surprising episode in which (almost immediately after learning about the *Same A-Same B* intuition rule) the prospective teachers falsely assert that a hexagon with all sides the same length must necessarily have all angles congruent, a claim that they continue to assert over an extended period of time, despite the failure of three different proof attempts. Tsamir (2007) uses this episode to illustrate the power of intuition and its resilience in the face of attempts to establish norms of logical reasoning. In particular she argues that explicit attention to "the fragility of their mathematical knowledge and the power of intuition" (p. 270) can support students in the development not only of pedagogical content knowledge (PCK) but also in their subject matter knowledge (SMK). Likewise, teacher education scholars (Steele, 2006; Steele and Hillen, 2012) have argued for incorporating a content dimension to pedagogy courses, and shown that some of those interventions increase the mathematical knowledge for teaching geometry. The possibility of considering mathematical content knowledge as part of what gets transacted in pedagogy classes has been supported by the development of new ways of thinking about mathematical knowledge for teaching. In the next section, we elaborate on this distinction between different domains of content knowledge.

4.3.2. The Role and Nature of Teachers' Knowledge in the Context of Geometry

As was noted in the prior section, preservice teacher education—and in particular, the explicit study of geometry by future teachers—is but one context in which it makes sense for researchers to examine the role and nature of geometry teachers' knowledge. In this section we examine the idea of geometry teachers' knowledge more expansively, situating it within a broader paradigm of research on teachers and teaching. We wish to understand:

1. What are the ways in which educational research has conceptualized teacher knowledge, and what methods have been developed by the field for studying teacher knowledge?
2. To what extent have those theories and methods been applied to the specific case of secondary geometry teachers?
3. What has been learned about secondary geometry teachers' knowledge, and how it interacts with their practice?

4.3.2.1. The Emergence of an Interest in What Teachers Know

The modern era of inquiry into the relationship between knowledge and teaching began with Shulman (1986a), who first articulated a viewpoint that has, in the ensuing three decades, become rather commonplace: That expert teaching draws on specialized forms of knowledge beyond knowledge of subject matter and general pedagogy. In examining the paradigms for research on teaching, Shulman (1986a) criticized the focus on general cognitive processes of then current research on teachers' cognitive processing of the act of teaching and claimed that we need an "elucidation of teachers' cognitive understanding of subject matter content and the relationships between such understanding and the instruction teachers provide for students" (p. 25). Shulman (1986b) distinguished between three categories of content knowledge: (a) subject matter content knowledge, (b) pedagogical content knowledge, and (c) curricular knowledge. The first of these categories includes not only "knowledge of the facts or concepts of a domain" but also "understanding the structures of the subject matter" (p. 9). Shulman (1986b) did not exemplify these distinctions in the specific case for mathematics but other scholars following in his footsteps have done that. Shulman's (1986b) distinction between knowledge of facts and knowledge of structure argues in particular for subject matter content knowledge in geometry to include not only the various properties of geometric figures, but also an understanding of the structure of geometry as a logico-deductive system, in which theorems are proven from axioms and postulates. The second of Shulman's categories, *pedagogical content knowledge* (or PCK, as it has subsequently come to be known), includes "the particular form of content knowledge that embodies the aspects of content most germane to its teachability" (p. 9). Shulman makes no attempt at a detailed map of this intellectual territory, but he does sketch out some of its contours:

> Within the category of pedagogical content knowledge I include, for the most regularly taught topics in one's subject area, the most useful forms of representation of those ideas, the most powerful analogies, illustrations, examples, explanations, and demonstrations—in a word, the ways of representing and formulating the subject that make it comprehensible to others. Since there are no single most powerful forms of representation, the teacher must have at hand a veritable armamentarium of alternative forms of representation, some of which derive from research whereas others originate in the wisdom of practice.

Teachers of geometry have access to a diverse and specific set of representations that help illustrate this point—for example, how to use the geoboard to explore the area of irregular figures (Moyer, 2001), or how to use dynamic geometry software to expand students' understanding of proof (de Villiers, 2004). In addition to speaking about the representations known to teachers, Shulman (1986b) includes among pedagogical content knowledge some understanding of students' learning:

> Pedagogical content knowledge also includes an understanding of what makes the learning of specific topics easy or difficult: the conceptions and preconceptions that students of different ages and backgrounds bring with them to the learning of those most frequently taught topics and lessons. If those preconceptions are misconceptions, which they so often are, teachers need knowledge of the strategies most likely to be fruitful in reorganizing the understanding of learners, because those learners are unlikely to appear before them as blank slates.
>
> *(Shulman, 1986b, pp. 9–10)*

Indeed some of the observations shared in Chapter 3 about students' thinking and learning of geometry are known to some teachers and used by them as they make instructional decisions.

Shulman describes the third of his knowledge categories, *curricular knowledge*, as an awareness of the range of materials available for instruction, including (for example) "alternative texts, software, programs, visual materials, single-concept films, laboratory demonstrations, or 'invitations to enquiry'", and the ability to make informed judgments about when and how to use them. Shulman also included in this category both *lateral curriculum knowledge* (knowledge of the curriculum materials being studied by students in other courses at the same time) and *vertical curriculum knowledge* (knowledge of what has been and will be taught in the same subject area in students' previous and future studies). Again there are notable elements of curricular knowledge in geometry, such as tools that can be used in the geometry course (e.g., the set square to draw perpendiculars), elements of lateral curriculum knowledge such as maps that can be used in geometry as well as in geography and history, and elements of vertical curriculum knowledge such as Cuisenaire rods and tangrams that could be reused or at least evoked to represent the ideas of secondary geometry.

It is in the identification of PCK that Shulman (1986b) has made the greatest impact—subsequent research on teachers and teacher knowledge has routinely and consistently invoked this construct in grounding further investigations and developing finer constructs. It is worth noting that Shulman's (1986b) epistemology of teacher knowledge does not stop with the identification and description of these three knowledge categories. Shulman also describes three *forms* of knowledge—*propositional knowledge, case knowledge,* and *strategic knowledge*—each of which is found in all three of the categories just discussed. Propositional knowledge is itself subdivided into three types of propositions, which Shulman calls *principles, maxims,* and *norms.* Case knowledge is divided into three types of cases, called *prototypes, precedents,* and *parables*. Finally, strategic knowledge consists of an awareness of how, when and why to draw upon all of the other forms of knowledge, how to reconcile conflicting propositions or seemingly contradictory cases, and how to communicate the reasons for his or her actions to others.

Shulman (1986b) launched an era of research into the nature of teacher knowledge and into how that knowledge grows and changes over time. Others have added new questions. For example, Fennema and Franke (1992) call attention to the *situated* or *contextual* nature of teachers' knowledge; that is, "within a given context, teachers' knowledge of content interacts with knowledge of pedagogy and students' cognitions and combines with beliefs to create a unique set of knowledge that drives classroom behavior" (p. 162). We come back to this point below when we connect teacher knowledge of geometry with the instructional situations that serve as contexts of work.

4.3.2.2. Knowledge Development: Connecting Opportunity to Learn and Knowledge Use

Shulman's epistemology of teacher knowledge has been the first step in a number of attempts to map the terrain of teacher knowledge; in mathematics education there have been numerous efforts, as can be gleaned from the review by Depaepe et al. (2013). Rowland and Ruthven (2011) and Blömeke and Delaney (2014) include elaborations of many of those perspectives, from theory to methodology. An important motivation in the progress of conceptions of teachers' knowledge of mathematics from earlier considerations of coursework in college to later considerations that tie it to pedagogy, curriculum, and students is the sense that what matters to students' learning is the knowledge used in, involved in, or called for in the practice of teaching, not merely the knowledge held by the teacher (Ball, et al., 2001).

Some researchers have been interested in mathematical knowledge for teaching from a developmental perspective (e.g., Silverman and Thompson, 2008). Rowland et al. (2005), in particular, have developed a framework they call the "knowledge quartet", as a means of both reflecting on and developing teacher knowledge. This framework emerged out of a study conducted in the UK, in which 24 mathematics lessons taught by trainee primary school teachers were analyzed using grounded theory methods, generating a set of 18 codes which were then aggregated into the four components of the knowledge quartet: *Foundation, Transformation, Connection,* and *Contingency*. These four components correspond, approximately, to lower or higher degrees of engagement with the moment-to-moment nature of teaching. The first component of the quartet, *Foundation*, describes teachers' learning at the university, in preparation for their work as teachers. Such knowledge is primarily (in Shulman's terms) *propositional* in nature. The second component of the quartet, *Transformation*, describes the capacity of teachers to transform their content knowledge into a lesson plan.

This component includes the ability to draw from print and online published resources, and the selection and use of examples for demonstration, group work, and individual student practice. The third component, *Connection*, concerns "the coherence of the planning or teaching displayed across an episode, lesson or series of lessons" (Rowland et al., 2005, p. 262). Finally, the fourth component, *Contingency*, concerns competence with classroom events that are (in the authors' terms) "almost impossible to plan for" (p. 263). This includes the ability to respond in real time to children's ideas and a preparedness to improvise. The developmental framework by Rowland et al. (2005) could be used to design studies that describe how different approaches to the geometry preparation of secondary teachers (e.g., courses on Non-Euclidean or differential geometry, courses in Euclidean geometry) enable their development of plans, lessons, and responses to students. In addition to the interest in understanding how preservice teachers develop geometry knowledge for teaching, researchers have been interested in elaborating Shulman's (1986b) ideas and defining more precisely what mathematical knowledge for teaching consists of. The next section discusses this.

4.3.2.3. Extending the Epistemology: Mathematical Knowledge for Teaching (MKT)

In recent years one of the most influential approaches to studying the nature of mathematical knowledge as it is applied to the work of teachers has been founded in the conceptualization by Ball et al. (2008). Ball et al. (2008) begin their conceptualization of Mathematical Knowledge for Teaching (MKT) with Shulman's distinction between Subject Matter Knowledge (SMK) and Pedagogical Content Knowledge (PCK) and undertake to make it more precise for the case of mathematics. Each of those subdivisions is divided into three domains: SMK into Common Content Knowledge (CCK), Specialized Content Knowledge (SCK), and Horizon Knowledge (HCK), and PCK into Knowledge of Content and Students (KCS), Knowledge of Content and Teaching (KCT), and Knowledge of Curriculum (KC). In addition to contributing a (sometimes confusing) array of two- and three-letter abbreviations to the lexicon of mathematics education, the MKT framework has been influential largely because Ball and her collaborators have been able to develop and validate items for measuring MKT at the elementary- and middle- school level. They have used these instruments to study more than 300 first- and third-grade teachers, and found that teachers' mathematical knowledge (as measured by the MKT items) was significantly related to student achievement gains (Hill et al. 2005); they have also found that teachers' MKT is related to the mathematical quality of their instruction (Hill et al., 2008).

Ball et al.'s (2008) elaboration of Shulman's categories has been useful for locating how research on cognition and learning fits into the larger picture of teachers' knowledge. Tabach et al. (2010) make a difference between two different aspects of teachers' subject matter knowledge: On the one hand they have "to produce solutions, strategies and explanations" just like their students will need to do; on the other hand they will have to appraise such solutions and strategies, a task of teaching. In doing such appraisal, Tsamir et al. (2008) note that teachers need what Ball et al. (2008) call KCS: In geometry this concerns children's ways to intuitively recognize examples and non-examples as well as knowledge of children's commonly held concept images and concept definitions for geometrical figures (Tall and Vinner, 1981). A question that remains for our field to investigate is, what among the research done on students' cognition and learning in geometry is actually useful for teachers of geometry to interpret and appraise their students' work?

Another approach that builds on the MKT work, but with quite a different focus, attends to what Ball et al. (2008) called "horizon content knowledge". Although the MKT research to date has largely refrained from elaborating on the nature of horizon knowledge, others have explored the nature of this particular facet of teacher knowledge. Zazkis and Mamolo (2011) argue that "teachers' horizon knowledge is. . . deeply connected to their knowledge of advanced (university or college level) mathematics". Elsewhere, Zazkis and Leikin (2010) refer to the "knowledge of the subject matter acquired during undergraduate studies at colleges or universities" as Advanced Mathematical Knowledge (AMK) and explore the ways in which such advanced knowledge can be instrumental for teachers engaged in the work of day-to-day teaching. Thus in seeking to understand the forms and nature of the knowledge that teachers can bring to bear in their practice, we find ourselves returning full circle to the need for a richer description of the mathematics content courses that teachers take as part of their preservice education, and to the importance of articulating the ways in which that mathematics content can be relevant for preservice teachers' future practice. Specifically, preservice teachers learn advanced mathematics in their teacher preparation. Some of that mathematics—e.g., abstract algebra, linear algebra, and differential geometry—provides important principles and ways of thinking that relate to the geometry that these teachers will teach in school. But how do those advanced mathematics fields support the geometry teacher's work with his or her students?

An important achievement of the MKT framework is the identification of the domain called SCK, specialized content knowledge, which has sometimes been defined as the knowledge of mathematics that only teachers have (a rather unfelicitous definition, in our view) and sometimes as the knowledge of mathematics that is needed to do the tasks of teaching. This latter definition provides an important opportunity to connect with the attention to the contexts of knowledge use called for by Fennema and Franke (1992). Indeed, one can anticipate that the tasks of teaching mathematics may vary depending on the age of students, and the mathematical work students will need to do to learn the course of studies they are offered. This has particular connections to geometry that could be investigated and that tie back to themes introduced before: When teachers draw diagrams for students to use in various tasks, they themselves need to do some mathematics, a mathematics that may stay hidden from students' view but that makes the students' work possible.

The construction of representations of mathematical ideas is a task of teaching that calls for such mathematical work. Adler and her collaborators (see e.g., Adler, 2005;

Adler and Davis, 2006) call particular attention to what Ball refers to as "unpacking" of mathematical knowledge (Ball and Bass, 2000; Ball et al., 2004). For Adler (2005), this "unpacking" provides "a compelling description of the distinctiveness of the mathematical work that teachers do", and she asks, "Where and how are prospective teachers provided with opportunities for learning to unpack mathematics?" (Adler, 2005, p. 5). Their exploration of inservice mathematics teachers documents how the practice of teaching creates opportunities for teachers to unpack mathematical ideas, suggesting that some elements of MKT are actually learned on the job. After introducing what has been done more generally to investigate teacher knowledge of geometry, we return to more recent investigations of MKT in secondary geometry

4.3.2.4. Mathematical Knowledge for Teaching Geometry

Despite the wealth of alternative approaches available for researchers who wish to attend to mathematics teachers' knowledge, there remains a surprising dearth of attention to the specific case of mathematical knowledge in the context of teaching geometry. Swafford et al. (1997) noted that while it has been demonstrated—for example, by the work of the Cognitively Guided Instruction project (e.g., Carpenter et al., 1989)—that research-based knowledge of student cognition is important for instructional practice in number and operations, "little is known about teachers' knowledge of student cognition in geometry or the impact of that knowledge on instruction" (Swafford et al., 1997, p. 468). In an attempt to fill that gap, Swafford, et al. (1997) studied 49 middle-grade teachers in the United States who participated in a 4-week program, consisting of a course in geometry and a research seminar on the van Hiele theory. Pretest-posttest results showed that participants in the program experienced increases in both content knowledge and van Hiele level, as measured by instruments adapted from the Cognitive Development and Achievement in Secondary School Geometry (CDASSG) project at the University of Chicago (Usiskin, 1982; see Chapter 3). More significantly, follow-up observations of eight teachers showed "marked changes" in their teaching, which the teachers ascribed to what they had learned from the course and research seminar. These changes included the addition of content areas not previously studied in their courses (including tessellations and star polygons), the inclusion of nonstandard tasks (including open-ended problem-based learning), and an increased focus on discourse and discussion:

> These patterns, based on teacher reports and teacher and researcher perceptions, suggest that following the geometry summer program, teachers (a) were spending more time and more quality time on geometry instruction; (b) were more willing to try new ideas and instructional approaches; (c) were more likely to engage in risk-taking that enhanced student learning; and (d) were more confident in their abilities to provoke and respond to higher levels of geometrical thinking. The teachers' reactions to the

> classroom episodes reveal that these instructional patterns were influenced by increased geometry content knowledge and research-based knowledge of student cognition.
>
> *(Swafford et al., 1997, p. 476)*

Participants also reported making deliberate attempts to design lessons and tasks at a higher van Hiele level than they otherwise would have.

Other scholars who venture into this relatively new field of research on teachers' knowledge of geometry have remarked on the lack of a well-established research base for understanding the interaction between teachers' knowledge and the teaching of Geometry—at the elementary and middle-grades levels as well as at the secondary level. Jones et al. (2002), discussing the state of primary teachers' knowledge of geometry for teaching in the UK, note that "the question of what trainee primary teachers need to know about geometry in order to teach the geometry component of the mathematics curriculum effectively and confidently is, at the moment, an open question" (Jones et al., 2002, p. 97). They further observe that the UK government regulations then in effect, issued in 1998, when compared to analogous recommendations for the mathematical education of teachers in the US by the Conference Board for the Mathematical Sciences (2000), differ primarily in that the US specification included "mathematical processes (understanding the role of mathematical definition; making conjectures and then proving or disproving; the ability to visualize and solve problems; the connections between mathematics and other subjects; and so on)" (Jones et al., 2002, p. 98), processes which were largely absent from the UK standards. Their review of the state of the research (as of 2002) concluded that knowledge-for-teaching-geometry was essentially unexplored territory, especially in the context of primary teachers.

In an attempt to fill this gap at the level of secondary geometry, Herbst and Kosko (2012, 2014) report on their development of an instrument for measuring mathematical knowledge for teaching high school geometry (MKT-G). Beginning with the model of MKT found in Ball et al. (2008), and following item development procedures analogous to those used by Ball, Hill and colleagues, Herbst and Kosko (2014) report on a project that developed 75 items, each related a priori to one of four domains of the MKT framework including CCK, KCS, KCT, and SCK.

To aid in generating items to measure knowledge of mathematics used in the tasks of teaching, Herbst and Kosko (2014) generated a list of tasks of teaching, such as "designing a problem or task to pose to students; evaluating students' constructed responses, particularly student-created definitions, explanations, arguments, and solutions to problems; creating an answer key or a rubric for a test; and translating students' mathematical statements into conventional vocabulary" (pp. 26–27). The tasks of teaching were particularly useful to create items that could arguably measure specialized content knowledge: They noted, however, that the nature of the mathematical work involved in such tasks of teaching varied, not only with the task (i.e., teachers do different mathematics when creating a problem and when interpreting students' solutions) but also with the specific *instructional situation* that frames the work (i.e., creating a diagram for an exploration requires different mathematical work than doing so for a calculation; see Herbst, 2010, and section 4.2.2.4 above).

Herbst and Kosko (2014) initially submitted their items to cognitive pretesting (Karabenick et al., 2007), in which they interviewed 11 teachers as they completed the 75 items. Teachers' responses to those items supported an examination of the content validity of the items and, in some cases, led to revision of the item wording. The revised set of items was then piloted with 48 participants. Herbst and Kosko (2014) used classical test theory to select a set of 34 items that had good statistical fit with the overall response pattern of the other items in each of the associated domains. These retained items were then piloted with a second pool of 35 additional teachers, for a combined sample of 83 participants. Individual scores were computed for each of the four domains, and moderate to strong correlations between the domains were found, indicating that "the different domains are, to a degree, interrelated" (Herbst and Kosko, 2014, p. 31). The pooled sample of 83 participants also made it possible to model the data using a one-parameter IRT (item response theory) model treating MKT-G as a single construct. Of this IRT modeling, Herbst and Kosko (2014) say

> Results from our IRT modeling show sufficient item reliability (0.96) and person reliability (0.82). Item reliability indicates how well our survey distinguishes between easier and more difficult items, and is generally considered acceptable if above 0.90. Person reliability indicates how well our survey distinguishes between groups of people (e.g., lower and higher scorers), and is generally considered acceptable if above 0.80.
>
> *(p. 33)*

Further analysis of the aggregated, scaled MKT-G scores showed that high scores were significantly correlated with experience teaching geometry, replicating an earlier result (Herbst and Kosko, 2012) that had found significant correlations between experience teaching geometry and raw scores but no correlation between experience teaching mathematics and raw scores. In contrast, Herbst and Kosko (2014) found no correlation between MKT-G scores and university-level course taking in mathematics in general or geometry in particular. These last findings are particularly interesting, especially in light of the studies we have summarized above regarding the intended curriculum of preservice teacher education in geometry; They suggest that the way teacher education programs conceptualize and attempt to teach geometry for teachers may not align well with the kind of knowledge that is actually salient for teaching geometry, and that this latter form of knowledge is typically learned on-the-job through the experience of teaching the subject.

Furthermore, from examining how performance varied between items, Herbst and Kosko (2014) noted that experienced geometry teachers did better than teachers without experience teaching geometry in items that were contextualized in an instructional situation. For example, in a problem that required respondents to choose a set of algebraic expressions for the sides of an isosceles triangle so that the students could be asked to do a geometric calculation, experienced geometry teachers performed much better than teachers without experience teaching geometry. But in problems that confronted participants with managing students' encounter with novel tasks, the performance of teachers with and without experience teaching geometry was comparable.

This item-by-item analysis suggested a possible reason why experience teaching geometry might increase specialized content knowledge for teaching geometry: As teachers recurrently engage in instructional situations, they get to solve similar kinds of mathematics problems with some frequency (e.g., they get to come up with expressions for sides or angles, so that students can engage in geometric calculation) and this practice increases their specialized content knowledge. This conjecture suggests also how we could move beyond the basic, intensive definition of specialized content knowledge for teaching geometry that defines SCK-G as the knowledge of mathematics involved in doing the tasks of teaching geometry. We could move to an extensive definition by identifying, for each instructional situation in geometry, the mathematical conceptions (i.e., problems, operators, representations, and controls; see Chapter 2) involved in doing the mathematics a teacher needs to do to perform the tasks of teaching called forth by the situation. Furthermore, because a situation is articulated by a system of norms, they provide a basis to judge the quality of the conceptions employed.

Consider as an example the case of a situation of exploration and let us speculate on what might be part of the specialized content knowledge a teacher needs to do their part well in this situation. Consider the property that the diagonals of a rectangle are congruent and let's imagine a teacher who wants to give students a rectangle and ask them to explore it, expecting that one of the properties they will state is that diagonals are congruent. The situation calls for the teacher to provide a diagram that students can use to conjecture a property; for this task of teaching, the teacher needs to draw a rectangle. There are several ways in which a rectangle could be drawn. Consider the following two possibilities:

(a) the teacher might draw a segment AB of a given length x, draw a perpendicular through B and measure on that perpendicular a length y, marking point C at that length, draw a perpendicular to BC through C, measure a length x on it, and mark point D at that length, and finally connect D with A.
(b) the teacher might draw a circle, then draw two lines that pass through the center of the circle, labeling the points at the intersections of each line and the circle, connecting those points, and then erasing the circle.

We should say that if both construction procedures were perfectly executed, both would achieve an actual rectangle (this is Common Content Knowledge of geometry). But what the teacher is doing is creating a diagram, using procedures that include random errors that are compounded when building on each other. There are several sources of error in procedure (a) including that the lines that are supposed to be perpendicular may only be very close to being perpendicular or that the segments that are supposed to be equal may only be very close to being equal. That error affects the determination of points C and D, which could result in the strokes that represent segments $\overline{AC}$ and $\overline{BD}$ not having equal length. On the other hand, the sources of error for procedure (b) seem less likely—if the compass opening does not remain constant or one of the lines does not pass through the center of the circle, the figure may turn out not to be a rectangle. Clearly both procedures (a) and (b) involve doing some mathematics as part of complying with a task of teaching, to draw a diagram for students to explore. We could say that each of them belongs to a conception of rectangle construction, and that in principle for that reason they are part of SCK. But since the instructional situation calls on students to interact proximally with the diagram and read off properties of the figure, we contend that procedure (b) is more likely to be correct than procedure (a). The reason is that procedure (a), being more likely to contain construction inaccuracies, could originate a drawing that appears like a breach of the norm: The students are expected to observe and verify that diagonals are congruent, but the diagram might not support that verification.

Obviously, we might question the extent to which the situation of exploration is mathematically productive. From the modeling perspective we espouse in this book, we might consider the breach alluded to above as a desirable action—aimed at having students confront an imperfect reality and try to muster the means to control it. Hence the importance of the construct *instructional situation* as a container for SCK: While procedure (b) might be the correct one to draw a rectangle for an exploration, procedure (a) might be the best one to appeal to if what the teacher is trying to do is to create a drawing of a rectangle that does contain inaccuracies. Back to Fennema and Franke's (1992) point, this illustrates the *situated* or *contextual* nature of teachers' knowledge. Indeed it illustrates how sociotechnical resources (in this case situational norms) interact with teacher knowledge.

The results found by Herbst and Kosko (2014) suggest that MKT-G might primarily be learned on the job, rather than through preservice education. That finding is consistent with critiques such as Wu's (2011) that point to the lack of enough attention to plane Euclidean geometry in university courses for teachers. However, researchers have also looked at preservice mathematics teacher education from other perspectives. The next section reports on what research has found and suggests places where new research can go.

4.4. Studies of Preservice Teachers' Knowledge of Geometry

There is not a lot of literature that accounts for preservice teachers' knowledge of geometry, let alone their knowledge of geometry for teaching. Mayberry's (1983) was among the first studies of preservice teachers' knowledge of geometry. As discussed in Chapter 3, Mayberry constructed a 128-item test intended to measure participants' van Hiele levels with respect to seven different topics in geometry (squares, right triangles, isosceles triangles, circles, parallel lines, similarity, and congruence). The test was administered to 19 preservice elementary teachers, of whom 13 had taken a geometry course in high school. Each of the 133 (= 19 × 7) combinations of student and topic yielded a *response pattern*, which Mayberry used to classify the students' van Hiele levels on a per-topic basis. The findings of the study were rather grim:

> There was a difference in levels achieved by those students who had taken high school geometry and those who had not. There were 68 response patterns with Level I or below as the highest level. Thirty (44%) of these were generated by the six students (31%) who had not taken a geometry course. However, of the 91 response patterns generated by the students who had taken geometry, only 30% were at Level III. . .
>
> (p. 67)

Of course, given that the participants in Mayberry's study were preservice elementary teachers, it is not at all clear that any conclusions may be drawn from it about the knowledge of future teachers of geometry at the secondary level. Perhaps surprisingly, the research literature contains very few analogous studies that focus on preservice secondary teachers' knowledge of geometry, whether measured using van Hiele levels or otherwise. One example of such a study is that of Halat (2008), who administered a 25-item van Hiele Geometry Test to both preservice elementary (N=125) and secondary (N=156) teachers at a university in central Turkey. The study found "no statistically significant difference in regard to the reasoning stages between the pre-service elementary school and secondary mathematics teachers". In particular, Halat (2008) found that only 6.4 percent of preservice secondary mathematics teachers scored at the two highest van Hiele levels, i.e. Level IV (Deduction) and Level V (Rigor), with more than half of all preservice secondary teachers (57.7 percent) at or below Level II (Analysis). Halat (2008) also found a statistically significant (p = .004) gender difference in the measured van Hiele levels of the preservice secondary teachers in the study, with males at a slightly higher mean van Hiele level (2.49) than females (2.07). It should be noted however that this might be a national particularity as results of the TEDS-M study suggest (Schmidt et al., 2011).

Similar findings were reported in an unpublished Masters thesis by Knight (2006), who measured the van Hiele level of preservice elementary and secondary teachers both before and after taking a course in geometry at the University of Maine. Knight (2006) found that the average score of the preservice secondary teachers in her sample was statistically significantly lower than Level IV (Deduction) both before taking a Higher Geometry course ($\overline{X} = 2.5$) and after taking the course ($\overline{X} = 3.125$).

But as we have seen in previous sections, measuring preservice secondary teachers' van Hiele levels is only one approach to studying their knowledge or understanding of geometry—and, given that the van Hiele theory was not conceptualized around the specialized nature of the mathematical work that teachers do, perhaps not the most salient. Furthermore, the critiques we have raised in Chapter 3 to the van Hiele theory are also pertinent when we discuss teachers' knowledge of geometry. For an example of quite a different approach, we consider the work of Chinnappan, Nason, and Lawson (1996), who propose a model of "the relationship between the organization and accessibility of teachers' subject-matter knowledge, the nature of their teaching, and the nature and quality of student learning" (p. 115). They use their model to structure a case study of a single pre-service secondary mathematics teacher's pedagogical and other content knowledge in the specific contexts of geometry and trigonometry. The pre-service teacher, referred to in the study by the pseudonym "David," participated in two interviews that focused on his knowledge of topics in those two contexts. The investigators found that David demonstrated "acceptable levels of [content] knowledge", but also that

> there was little evidence that the knowledge components were connected in a manner that would lead to the establishment of more extended networks of mathematical knowledge that are argued to be characteristic of expert teachers' mathematical knowledge base. . . There was little evidence that the student teacher had considered the possibility of reconceptualising or restructuring the mathematical content knowledge in preparation for presentation to students. . . Although there are instances where he had shown consideration to the learner in terms of how he or she would handle the concepts, there is a lack of understanding about how students activate and exploit mathematical knowledge acquired in the classroom.
>
> *(p. 121)*

Of course, one could look at teacher cognition from a conceptions (spheres of practice; Balacheff and Gaudin, 2010) perspective as well. That perspective would recommend looking beyond mental representations and also into the tools and signs used in engaging in problems; with such attention to knowing in practice, it might be possible to see connections and networks that are not yet visible when knowing is reduced to mental representations.

4.5. Another Look at Elementary and Middle Grades Teachers

At the end of section 4.3 we noted that very little research has been done on inservice secondary teachers' understanding of geometry; the reader may have noticed that the review of research on preservice secondary teachers in section 4.4 does not go much deeper, consisting as it does solely of two published papers and one unpublished Masters thesis. This scant knowledge base reflects the lack of scholarly attention that has, to date, been directed toward understanding secondary teachers' knowledge and understanding of geometry. This may in part be because a conceptualization of what it means to know geometry for teaching at the secondary level is still at a fairly early stage (see section 4.3.2.4. above). In contrast, there is an abundance of studies of preservice and inservice elementary and middle-grade teachers. Below, we briefly summarize four such studies. We caution the reader that it is not immediately clear in what ways, or even if, these studies of elementary and middle-grades teachers can inform our understanding of those who prepare to teach at the secondary level; nevertheless, we include those studies here partly for the international perspective they offer (the studies were based in Spain, South Africa, Turkey, and Indonesia), but primarily because the research questions they ask and the methods they employ may be useful for suggesting possible studies of preservice secondary teachers.

4.5.1. Prospective Primary Teachers' Memories and Conceptions of Geometry Instruction

Barrantes et al. (2006) focused on a group of prospective primary school teachers in Spain and studied the interaction between their explicit memories of their own experiences as students of geometry, and their conceptions about how geometry is or should be taught. They found (perhaps not surprisingly) that most participants primarily remembered their own experiences as students of geometry as being about memorizing and using formulas (e.g., formulas for areas of various geometric figures) without any explanation, derivation or justification of those formulas. Participants in the study typically did not distinguish between "problems" and "exercises"; that is, the phrase "problem-solving" meant for them an application of a formula to a standard-template exercise. The prospective teachers recalled, as students, having used manipulatives (models of geometric shapes, etc.), but that the use of these objects was primarily motivational; that is, "they recalled that they manipulated geometrical bodies made of wood, or constructed them out of card, but that they then carried out no activities with them" (p. 426). They conclude that primary school teachers in Spain begin teacher education "with almost no knowledge of geometry and with practically no referent as to how it is taught and learned," and that their conceptions are "far from the line of current guidelines" for how geometry should be taught.

4.5.2. Primary Teachers' Knowledge of Geometry for Teaching

Working in the context of the efforts to reform the Indonesian educational system, Ng (2011) studied how teachers' performance on the geometry component of the MKT instrument developed by Ball et al. (2008) was associated with personal characteristics such as school type (public or private), educational background (i.e., type of degree awarded), teaching experience, etc. The portion of the MKT instrument used in the study consisted of 19 questions aligned with geometry content commonly taught in grades 3–8. Ng (2011) translated the instrument into Indonesian and administered it to a sample of 167 primary teachers, finding that their mean MKT geometry score was −0.731 standard deviation units; that is, on average the teachers answered only 9 out of 19 questions correct. Ng found that MKT in the sample was significantly associated with educational background ($p < .001$); because teachers in Indonesian private schools on average have stronger educational backgrounds than those in public schools, this in turn led to an association between school types and MKT. One of the more interesting findings of the study was that teachers with experience teaching a wider range of grade levels outperformed teachers with experience in a more narrow grade band, regardless of whether that grade band was at the lower primary (1–4) or upper primary (5–6) level. On the basis of this finding Ng speculated that teaching a wider range of grades may help to develop a teacher's understanding of the structure of the mathematical content and how topics are sequenced in the curriculum; in the language of the MKT model, having experience teaching a wider range of grades contributes both to Knowledge of Content and Curriculum and to Horizon Content Knowledge.

4.5.3. Prospective Middle Grades Teachers' Understanding of Geometry

In a study based in South Africa, van der Sandt and Nieuwoudt (2003, 2005) studied a group of 100 preservice mathematics teachers and 18 inservice grade 7 mathematics teachers using a modified version of Mayberry's test of participants' van Hiele level and the acquisition scales of Gutierrez et al. (1991; see Chapter 3). The findings of the study were in many respects disheartening; they found that "neither group had achieved a complete degree of acquisition (85 percent), even for the first level of geometric thought" (2003, p. 201) and concluded that "this could suggest that pre-service teacher education does not adequately prepare teachers for teaching geometry" (p. 203). In fact, their study suggested that, perhaps paradoxically, pre-service training had a small but negative impact on teachers' measured van Hiele levels; preservice teachers at the beginning of their post-secondary education scored higher on the Mayberry test than did preservice teachers in the middle of and at the end of their education. One of the more interesting findings in their study was that inservice teachers consistently outperformed preservice teachers at the end of their formal education. They interpret this finding as suggesting that teachers' knowledge of geometry continues to grow over time; that is, teachers are learning through teaching. (This finding is consistent with the work of Herbst and Kosko, 2014, described in section 4.3 above.) Despite this, they concluded that neither preservice teachers nor their more experienced counterparts in the profession have mastery of the subject matter that they are supposed to teach, and called for a greater focus on preservice teacher education, especially focused on the content knowledge for teaching.

4.5.4. Prospective Middle School Teachers' Conceptions of Translations

Yanik (2011) studied 44 prospective middle school mathematics teachers in a large urban public university in central Turkey. In this study, Yanik (2011) used thirteen tasks in clinical interviews to probe participants' understandings of geometric translations. The interviews revealed that "the majority of participants had difficulty describing translations" and that nearly all participants' conceptions of translations were expressed not in terms of a rigid mapping of the plane onto itself but rather in terms of the physical motion of a single object. Strikingly, 45 percent of participants described translation in terms as a form of rotation, using the language of an object rolling from one position to another. Of those participants, half explicitly rejected the possibility that a sliding, non-rolling object could be considered an example of a translation. Participants who regarded translation as a form of rotational motion correctly held that the overall size and shape of an object were preserved under translation, but incorrectly held that the relative positions of the points within that object (e.g. the vertices of a polygon) were not necessarily invariant under a translation.

4.6. Beliefs of Secondary Geometry Teachers

In the previous sections, we have considered what is and is not known about the nature and structure of secondary teachers' knowledge of geometry, and how that knowledge develops along the preservice-inservice career path. But knowledge, no matter how important it is, is only one of the many personal resources that teachers bring with them into the classroom and that influence their practice. To complement studies of knowledge, we can also consider teachers' beliefs and attitudes, and how those affect their practices.

As is the case with research on teacher knowledge, there is a long tradition of scholarly inquiry into these dispositional variables in mathematics education. Beginning in the 1970s (Aiken, 1970, 1976), one major field of inquiry has been the study of teachers' attitudes, beliefs, and anxiety towards mathematics, and how these influence student learning. This interest led to the development of a number of instruments for measuring attitudes, most notably the Fennema-Sherman Mathematics Attitude Scales (FSMAS) (Fennema and Sherman, 1976; Mulhern and Rae, 1998) and various versions of the Mathematics Anxiety Rating Scale (Richardson and Suinn, 1972; Plake and Parker, 1982), which have been used extensively over the past 40 years. Some research done in Turkey has explored instruments to measure attitudes (the "Geometry attitude scale"; Bindak 2004) and self-efficacy (the "Self-efficacy scale towards geometry"; Cantürk-Günhand and Baser, 2007) in geometry. Furthermore, Ünlü et al. (2010) found a strong correlation between the scores on attitudes and on self-efficacy in geometry. Similar results are known for mathematical problem-solving with high school students (Pajares and Miller, 1994) and for general mathematics teachers in mathematics in South Africa by Nkwe (1985). In recent years the research focus on *attitudes* has gradually been displaced by a focus on *beliefs* (e.g., the Mathematics Teaching Efficacy Beliefs Instrument (MTEBI) and the Mathematics Beliefs Instrument (MBI; Carter and Norwood, 1997; Hart, 2002; Pajares, 1992; Stipek et al., 2001;Wilkins and Brand, 2004). More recent scholarship has provided ways to inquire into geometry-specific beliefs.

4.6.1. Beliefs on the Nature of Geometric Objects

Different frameworks for the description of teachers' attitudes on the teaching and learning of mathematics have been worked out in the context of both philosophy of mathematics and mathematics education. The Field medalist Réné Thom identified "the development of meaning, of the 'existence' of mathematical objects" (Thom, 1973, p. 202) as a crucial task for teaching mathematics. This goal suggests the need to ask what kind of object teachers think the geometric figure is. Girnat (2011) built on Kline's (1983) distinction between the formalistic, idealistic, and applied ontological views on geometric figures to study the ontological beliefs of geometry held by secondary teachers. That three-way distinction illustrates two considerations made by Stegmüller (1985) in his structuralist theory of science, according to which theories are characterized by the sets of axioms that establish their propositional content and by the sets of intended applications that establish their reference. *Formalistic*, *idealistic*, and *rationalistic* views derive their content from axiomatic geometry but differ in their reference: While the formalistic view denies the need of reference, the idealistic view identifies a world of ideal objects as the reference, and the rationalistic view refers to a world of physical objects. Girnat (2011) adds a fourth view which he calls *empiristic* and which is distinguished from the other three in that its propositional content does not come from axiomatic Euclidean geometry but rather from "geometrical assumptions . . . at most locally ordered, fulfilling the inferential standards of everyday discussions" (Girnat, 2011, p. 39). The empiristic view shares with the rationalistic view its reference to the physical world but derives its content from local considerations abstracted from intuition and experience. Girnat (2011) found that teachers' views of the geometric objects can be described using these four views of geometric objects.

Similar considerations were present in the distinctions made by Houdement and Kuzniak (2006) who identified three paradigmatic views on geometry: Natural geometry (Geometry I), natural axiomatic geometry (Geometry II), and formalist axiomatic geometry (Geometry III). The first geometry has as its purpose the validation of reality, but it already represents an effort of abstraction from reality, since thought chooses to excise certain aspects of objects if they are material or translates them into patterns. In the second geometry the source of validation is based on the hypothetical laws in a deductive axiomatic system, but it dwells in the problem of choosing axioms inasmuch as this geometry seeks to relate to reality and is organized according to geometric spatial problems. The third geometry is not bound to describing any pre-existing reality but rather it creates its world of objects through logical reasoning from abstract axioms (Houdement and Kuzniak, 2003, 2006). Kuzniak and Rauscher (2011) utilized these distinctions to investigate how teachers make sense of students' difficulties in a problem proposed in Geometry II, specifically noting that while some teachers recognized that students' difficulties might arise from tensions between Geometry I and Geometry II, others were not aware of Geometry I as a source of challenges for students.

Eichler and Erens (2015) report on a qualitative study in which the belief systems of secondary mathematics teachers were examined in domain-specific terms.

These scholars utilized a set of mathematical worldviews proposed by Grigutsch, Raatz, and Törner (1998) that distinguishes a formalist (world) "that stresses that mathematics is characterized by a strongly logical and formal approach" (Grigutsch et al., 1998, p. 184), a process-oriented view that emphasizes the heuristic and creative aspects of mathematical activity, an instrumentalist view that emphasizes the role of tools such as formulas and procedures to produce needed information, and an application-oriented view "that accentuates the utility of mathematics for the real world and the attempts to include real-world problems into mathematics classrooms" (Grigutsch et al., 1998, p. 184). Teachers of different mathematical subdomains were found to have different belief systems. In particular, while teachers of stochastics were found to be more aligned with the application view, geometry teachers were found to be more aligned with the process-oriented view.

Because teachers are ultimately responsible for enacting the intended geometry curriculum, any attempt to reform or modernize that curriculum must necessarily contend with teachers' beliefs and attitudes. One such reform attempt was undertaken in Iran in the mid-1990s. Gooya (2007) describes the former Iranian curriculum as one in which abstract deductive reasoning and two-column proofs predominated; the reform curriculum, in contrast, emphasized real-world applications, both inductive and deductive reasoning, and experimental approaches. To support the rollout of the new curriculum, its authors implemented a professional development program for 350 novice teachers and 350 veterans, among whom were 30 who were considered geometry specialists, teachers who had exclusively taught geometry for many years. Gooya (2007) studied these teachers' beliefs about the new curriculum, with particular attention to the divisions among "traditionalists", "incrementalists", and "innovators". Many of the beliefs teachers expressed concerned pedagogy, evaluation, and other issues that pertain to mathematics and its teaching in general. Some, however, expressed strong opinions about the essential nature of geometry in particular (and the fidelity of the new curriculum to that nature). For example, reacting to the de-emphasis of formal deduction and the new focus on experimental and inductive methods, one veteran teacher wrote that "Euclid's soul [would] be shivering in the grave from this dishonesty to his geometry!" (Gooya, 2007, p. 339). Other teachers were more receptive to the proposition of reconstructing the geometry curriculum; one novice teacher wrote that

> I should say to the old teachers who are worried about the declining status of geometry that geometry has not lost its status and dignity! However, it might be true that the status of those respected and experienced teachers *specialized in geometry* will be threatened! Those who have memorized every detail and have taught geometry for years without any reflection on what they are doing! Yes! They are threatened, because they have to study the new content, context, and approaches (emphasis in original).
>
> *(Gooya, 2007, p. 341)*

Reactions to the reform effort in Iran were reflective of, and interconnected with, broader issues of Iranian society. Gooya (2007) notes, for example, that while older mathematics teachers in Iran are almost exclusively male, about half of the participants in their professional development program were female, and female teachers (many of whom brought their children to the sessions) were the majority in the subgroup of novice teachers. These gender dynamics played out in how some of the participants related to the new geometry curriculum; one experienced teacher ascribed the curriculum changes to the authors' "motherly affection and disposition", arguing "wouldn't it be better if we look at the future from a fatherly perspective?" (Gooya, 2007, p. 341). Gooya (2007) concludes that in Iran, geometry "is losing its status as a purely deductive and male subject. It is gaining a different status by becoming more accessible to the majority of teachers and students, thus becoming a tool to enhance society" (p. 345).

4.6.2. Teachers' Beliefs about the Role of Proof in Geometry (and the Role of Geometry in Teaching Proof)

One persistent focus of the research into teachers' beliefs has been the role of proof and reasoning in mathematics. Since at least the 1990s, researchers in the United States have called attention to the fact that for many teachers proving is an activity that is strongly (and almost exclusively) associated with secondary geometry (Knuth, 2002; Schoenfeld, 1994; Wu, 1996). Moreover, Herbst (2002a) has documented how, in the United States, proving in secondary geometry has come to be enacted primarily through the ritualized form of the so-called *two-column proof.* Advocates for infusing proof and reasoning more broadly into the mathematics curriculum have thus called for de-emphasizing the two-column form in secondary geometry (NCTM, 1989) and simultaneously for seeking opportunities to engage in less formal modes of proof in other and earlier courses. For example, Harel and Sowder (1998) argued that "to delay exposure to reason giving until the secondary-school geometry course and to then expect an instant appreciation for more sophisticated mathematical justifications is an unreasonable expectation" (p. 674).

Although much of the literature on proof and reasoning focuses on teachers' and students' knowledge of proof—e.g., their ability to distinguish valid arguments from invalid ones—a smaller body of work focuses on teachers' attitudes toward proof. Mingus and Grassl (1999) found that an overwhelming majority (69 percent) of the preservice teachers they surveyed ($N = 51$) felt that proof should be introduced at an earlier stage, with many expressing the belief that the secondary geometry course was too late for students to have their first exposure to mathematical proof (p. 440). In contrast, Knuth (2002b) found that the majority of the secondary mathematics teachers in his study of secondary teachers' beliefs felt that proof was not appropriate for all students, and that it should be limited only to students in advanced mathematics classes (p. 73). For many of these teachers, the word "proof" was understood to be synonymous with the two-column form. Knuth (2002b) notes that these beliefs are fundamentally incompatible with the recommendations of reform documents such as the 2000 NCTM *Standards* and of mathematics educators such as Hanna (1983). Even among those teachers who granted that proof should have a role in higher-level secondary mathematics courses, most believed that this role was primarily implicit; geometry was the only course in which most teachers believed proof should play an explicit role (see also Knuth, 2002a).

Teachers' beliefs about proof and the role they play in shaping how teachers adopt reforms in the geometry curriculum were also considered in Laborde's (2002) study of the integration of dynamic geometry software into the French secondary curriculum. Laborde (2002) describes a group of four teachers (two experienced and two inexperienced) who participated in a three-year project dedicated to developing classroom tasks that made use of Cabri Geometry on hand-held TI 92 calculators. The tasks were piloted in the teachers' own classrooms, discussed with the full group, and modified for subsequent reuse; in some cases the process of modification and piloting was iterated multiple times over the full three-year span of the project. Laborde (2002) provides a thorough analysis of the different ways in which the participating teachers made use of technology and integrated dynamic geometry into their classrooms, and uses that analysis as grounds for formulating hypotheses about the teachers' beliefs about mathematics and their conceptions of teaching. Among these, Laborde (2002) identifies a "dichotomy between conjecture and explanation or proof. . . It was as if the process of elaborating a proof should deal with theoretical objects unrelated to their representations, not modified by actions on these representations" (p. 306). This orientation toward proof as an essentially abstract activity that must be disconnected from representations seems similar to the beliefs of the traditionalist teachers in Gooya's (2007) study.

Whereas this belief about the nature of proof was shared by all four participants in the study, Laborde (2002) also identified additional beliefs that were characteristic of the experienced or of the novice teachers. For example, the novice teachers believed that geometry is "intrinsically linked to paper and pencil" (p. 307) and that "technology is an additional component of the teaching system but external to the learning processes" (p. 310). The experienced teachers, in contrast, designed scenarios that "offered more open exploration activities" that were integral to the learning process. Even these teachers, however, "expressed clearly that the drag mode was for them more to facilitate visualization than to act in the solution process, even for the open-ended problem" (Laborde, 2002, pp. 310–311). Laborde (2002) speculates that teachers are slow to integrate technology into the curriculum not only because "it takes time for them to accept that learning might occur in computer-based situations without reference to a paper-and-pencil environment", but also because it "takes time for them to accept that they might lose part of their control over what students do" (p. 311).

4.6.3. The Practical Rationality of Geometry Instruction

In the discussion above, we have approached beliefs and attitudes from the perspective of individual teachers; that is to say, we have taken the stance that beliefs and attitudes, like knowledge, are qualities that individuals hold in idiosyncratic ways. But there is another fruitful perspective on beliefs that may be considered: In addition to the beliefs of individuals, we may inquire after the beliefs or dispositions available to members of a profession.

Teachers are, of course, individuals, with their own agency and volition. But as members of a practice, teachers are not entirely free to act in any way they wish; rather their work, like the activities of any social group, is subject to a practical rationality or habitus (Bourdieu, 1991): A complex network of sensibilities that are felt as personal but are also shared and used to hold others tacitly accountable. Herbst and Chazan (2003) have used the expression *practical rationality* to name this set of sensibilities. Some of those sensibilities are the norms of instructional contracts and situations and regulate what is expected to be done, by whom, at a particular moment. The norms of instructional situations and contracts provide regularity to teachers' practice; this is one of the reasons why, for example, a high school geometry class in one part of the US generally looks and operates much the same as a geometry classroom in another part of the US (but different from an algebra classroom in the same school, even one taught by the same teacher).

Despite their regularity, these norms are not prescriptive; rather they are a description made by an observer of what goes without saying in mathematics classrooms. Teachers can, and frequently do, depart from these norms, but when they do so they typically mark those departures, simultaneously confirming the norms and demonstrating their malleability. For example, Weiss et al. (2009) investigated secondary geometry teachers' willingness to depart from one of the norms that describes how the two-column proof form is normally used; namely, the norm that the argument recorded in a two-column proof is generated as an alternating sequence of *statements* and *reasons*, and that the temporal sequence with which these are generated corresponds precisely to the spatial form of the proof (Herbst et al., 2009). In a virtual adaptation of the ethnomethodological tradition of *breaching experiments* (Garfinkel & Sacks, 1970; Mehan & Wood, 1975), teachers were presented with a videotaped geometry lesson in which that norm was breached by a teacher who allowed a student engaged in writing a two-column proof to assume the truth of a statement without supplying a justification, provided he returned to complete the argument later. Teachers responses to the episode varied: Some argued that the teacher was wrong to allow that move, because the nature of proof itself requires that each step be justified before one may proceed to the next step; others argued that the teacher's decision was a legitimate one, because real mathematicians often work in precisely this fashion. What is significant here is that although the teachers disagreed about the particulars of the case, they were in agreement that the activity of proving in secondary geometry ought to be authentic to the work of proving among mathematicians. Weiss, et al. (2009) concluded that this imperative to make school mathematics authentic constitutes one of the shared values of the practice, what might be described as a professional obligation or group belief that individual practitioners position themselves relative to and invoke to justify their instructional decisions. Nachlieli and Herbst (2009) uncovered yet other such justifications that practitioners brought up to account for the same breach of an instructional norm.

To formalize the sense that there are common grounds upon which members of the mathematics teaching profession justify teaching decisions, Herbst and Chazan (2012; see also Chazan et al., 2016; Herbst & Balacheff, 2009) have identified four *professional obligations* that provide grounds on which a teacher might warrant actions and decisions. The four obligations are:

- The *disciplinary* obligation, the imperative to faithfully represent the practices and content of the discipline of mathematics;
- The *individual* obligation, the imperative to attend to the well-being of each student, including students' "behavioral, cognitive, emotional, or social needs" (Herbst & Chazan, 2011, p. 450);
- The *interpersonal* obligation, the imperative to shepherd social and societal values as a teacher manages interactions among students and shared resources in the classroom; and
- The *institutional* obligation, the imperative to conform to and comply with all aspects of the educational institution, such as "attending to school policies, calendars, schedules, examinations, curriculum, extra curricular activities" (Herbst & Chazan, 2011, p. 451).

These obligations, it should be noted, are understood to apply to mathematics teachers at all levels. In the particular case of secondary geometry, Herbst and his collaborators have investigated a broad range of norms that characterize instructional situations typical in the United States. These include:

- how theorems are "installed" (i.e., made part of the classroom discourse) in classrooms (Herbst et al., 2011), and in particular when such "installation" does or does not require that the theorem be proved (Herbst & Miyakawa, 2008);
- the extent to which teachers view themselves as obligated to lay emphasis on the logico-deductive structure of geometry, and in particular the different roles played in that structure by postulates and theorems (Weiss & Herbst, 2015);
- the different shares of labor that are apportioned to teachers and students when in the situation of "doing proofs" (Herbst & Brach, 2006);
- the expectations regarding the register in which proof problems are posed (Herbst et al., 2013);
- the semiotic resources used in geometry diagrams (Dimmel & Herbst, 2015a);
- the expectations of students' presentation of proofs (Dimmel & Herbst, 2015b); and
- the norms of the situation of geometric calculation in algebra (Boileau & Herbst, 2015).

Studies such as these provide a means for discussing teaching from a point of view that does not reduce all considerations to a description of the individual teacher's personal attributes, such as his or her knowledge, understanding, beliefs and attitudes. Rather, this perspective allows us to consider the individual as a participant in a collective activity, whose encounter with that activity is experienced through categories of perception and appreciation that are shared with their co-practitioners, and whose individuality becomes manifest in the enactments of (and departures from) those collective norms.

4.7. Conclusion

This chapter has presented a consideration of the teaching of secondary geometry from the perspective of the resources, individual and sociotechnical, available to enact this teaching. But how does the chapter fit into the purpose of the book, to promote a modeling approach in the teaching and learning of geometry—the study of the means to represent, control, and use geometric models, such as diagrams, to make sense of people's experience with shape and space? While this connection has been muted through the chapter we sketch here how this chapter assists that purpose as we move towards the next chapter where the matter becomes more apparent.

If the study of geometry in secondary schools is going to engage students in the development and improvement of models to understand experience with shape and space, teaching work will be needed to support such study. The reasons for this were outlined at the beginning of this chapter, when we described the essential role of the teacher in enabling students' mathematical work and exchanging this work for claims on the subject to be learned. A modeling approach to the study of geometry needs, beyond a curriculum, skillful management of those instructional exchanges. The likelihood that this support will be available depends on capacities both of the individual teachers involved and of the sociotechnical environments where these individuals work.

The focus on sociotechnical resources took us to identify interactional structures that we call instructional situations, and that presently articulate the work of teaching. While those interactional structures don't necessarily support a modeling approach to the study of geometry at present, they represent the background against which modified or redesigned interactional structures will need to deploy. Situations of construction, exploration, calculation, and proof provide a basis for design research that can look into the conditions of possibility for engaging students in modeling work. It is essential for our field to study how the interactional structures of geometry instruction can be modified to make it possible for students to engage in modeling. The next chapter will engage us in such inquiry.

The focus on individual resources took us to review the state of teacher knowledge and teacher beliefs both in practicing teachers and in preservice teachers. Frameworks, as well as some results, have been presented. On the one hand the research on teachers' knowledge, using cognitive frameworks similar to those used to investigate students' cognition, shows that teachers themselves vary quite a bit in their knowledge of geometry, though studies of mathematical knowledge for teaching suggest that experience teaching correlates with increases in MKT (Herbst and Kosko, 2014). The beliefs research also shows great variation, with representation of positions that still see geometry as valuable insofar as it is an instance of a formal mathematical system. Clearly, the emergence of new technologies and the needs of visualization have brought new contexts to appreciate geometric knowledge, but at any rate it seems that especially designed activities and programs for teachers to learn and to value geometry from a modeling perspective may be needed. While this is not a topic that we address in this book at any length, we offer it as a possible project to undertake later: What is the mathematical knowledge for teaching geometry that would be needed to teach geometry from a modeling perspective? What would be items that represent this knowledge and how does that knowledge relate to the mathematical knowledge for teaching needed to enact current instructional situations? Further, how could teacher education experiences promote the learning of mathematical knowledge for teaching geometry from a modeling perspective?

Notes

1 By sociotechnical resources we mean resources made up of social interaction and technical or technological artifacts (including specialized tools and signs).
2 This, by the way, illustrates how the work of teaching geometry is subject specific—all teachers create assignments for their students, but geometry teachers who create these kind of assignments need to worry about the consistency of information provided with multiple representations such as the diagram and the algebraic expressions, and this worry about consistency requires mathematical values as well as actual mathematical work.
3 On a scale from 1 to 5, mean = 3.66 and SD = 1.12.
4 On a scale from 1 to 5, mean = 4.04 and SD = 0.84.

5

IMPROVING THE TEACHING AND LEARNING OF GEOMETRY IN SECONDARY SCHOOL CLASSROOMS

5.1. Introduction

In Chapter 2 we put forward a proposal for conceiving of the teaching and learning of geometry as the study of models with which students predict and control their experiences with shape and space. Chapter 3 reviewed the literature and theory on cognition and learning that might support this proposal. Chapter 4 reviewed aspects of teaching and teacher knowledge that are germane to these issues. In this chapter we bring those pieces together to actually envision what the teaching and learning of geometry could look like.

In this chapter we integrate ideas of curriculum, learning, and teaching into a consideration of instructional improvement in geometry, particularly providing a way to address the following question:

> How can instruction be organized to facilitate students' study of the means of predicting and controlling their interactions with figures, where figures can be navigations of space, captures of large objects, descriptions of artifacts or diagrams, or constructions of artifacts or diagrams?

The question prompts consideration of some instructional designs, either observed in ordinary classrooms, documented from design research, or envisioned by us for the reader's inquiry, in which students (may) have the opportunity to increase their intellectual means of predicting and controlling their interactions with representations of figures. We examine the space of possibilities and pose some research questions that Masters students or practicing teachers doing research on their practice could use in the context of further developing and enacting those interventions. We begin, however, with some framing considerations for the study of geometry instruction that can help bring up those research questions eventually.

5.2. Communication Tasks: A Contribution of the Theory of Didactical Situations to the Design of Interventions

The theory of didactical situations (TDS; Brousseau, 1997) provides a means for thinking about instruction that seeks the development of targeted conceptions. We review it here with the goal of examining how the four kinds of conceptions of figure identified in Chapter 2 could be made to evolve. We present Brousseau's theory as describing a particular case of the theory of instructional exchanges (a decision that, we concede, may seem questionable to some, but that makes sense to us). As we have noted, instructional exchanges, from the teacher's perspective, involve carving a share of labor within which the teacher can observe the handling of an item of mathematical knowledge. In particular, when thinking about academic tasks (that is, of work that students play a central role in completing), instructional exchanges confront the teacher with the problem of making a symbolic exchange between (on the one hand) the moment-to-moment work the students have done on a task and (on the other hand) some discrete elements of the knowledge at stake. While the tasks students do are likely to respond to a learning design (e.g., have been designed so as to enable students to acquire or otherwise show evidence of knowing the knowledge at stake), as they are experienced by students these tasks might or might not reveal their instructional aim. In general, one could describe these tasks as confronting the cognitive agent (the student, the group of students) against a counterpart system that supports and receives the actions of the former and, in doing so, provides feedback that often informs the next actions of the cognitive agent. Brousseau (1997) calls this counterpart system the "milieu".

We have referred to the milieu, this key concept in Brousseau's (1997) TDS, in our earlier discussion of conceptions. As noted in Chapter 2, the milieu is the counterpart of the learner or actor in a mathematical task,[1] the system that receives the actions of the learner and that produces feedback on those actions. In Brousseau's TDS, there is interest in whether the milieu is a-didactical; that is, whether the way in which the milieu supports and feeds back into the action of the cognitive agent can be construed by the learners as devoid of didactical intention. To understand the notion of adidacticity, contrast two possibilities for a problem that involves students enlarging the pieces of a jigsaw-type puzzle. In the puzzle problem, one piece of the puzzle is a 4×10 rectangle, and the problem is to indicate how to enlarge all the pieces of the puzzle, given that this rectangular puzzle piece is to be modified so that the side which is originally 4 units long becomes 7 units long (this discussion is inspired by Brousseau's, 1997, p. 177, "puzzle" problem). The question to answer is how long the side that originally measured 10 should be when enlarged. (The correct answer, 17.5, results from multiplying 10 by $\frac{7}{4}$.) Students have been observed making various kinds of transformations, including adding 3 (which would make the long side of the rectangle 13) and multiplying by 2 and subtracting 1 (which would make the long side of the rectangle 19). We bring this task up to illustrate what it means for the milieu to be adidactical. Clearly, the teacher could provide feedback to students who answer 13 or 19 by saying that the enlargement prescribed will not keep the same shape. Alternatively the teacher could let the students continue enlarging other of the puzzle pieces, using the rule they proposed and letting them see whether the enlarged puzzle can be put together to look like the original one, which would be impossible in the cases just described. In the case in which the teacher lets students continue to enlarge other pieces, the lack of resemblance of the enlarged puzzled pieces with the original puzzle pieces might become apparent without the teacher's active participation. In this second case, the teacher is not the source of the feedback; rather, the feedback comes from the milieu itself. In this case one can say that the milieu is a-didactical, even though both the design of the pieces' original dimensions and the design of the question itself are such that they permit both the emergence and the repeal of those misconceived solutions (and this does attest to design by the teacher). As far as the validation of the student's action, because this is done by the material of the milieu (Margolinas, 1993) and not by the overt action of the teacher, it is called a-didactical.

In some cases the milieu may include other people, but it may still be an adidactical milieu as long as their actions can be perceived as unrelated to the teacher's intention. In communication tasks, the learner is positioned relative to a milieu of communication, against which he or she tests the quality of messages he or she issues; that milieu of communication could include other students acting against a milieu of action (see Figure 5.1). To provide a quick canonical illustration of the difference between an action task and a communication task, consider for a moment the classical "Race to 20" example (Brousseau, 1997, p. 21). Race to 20 is a two-person game in which one person starts by saying 1 or 2; the opponent

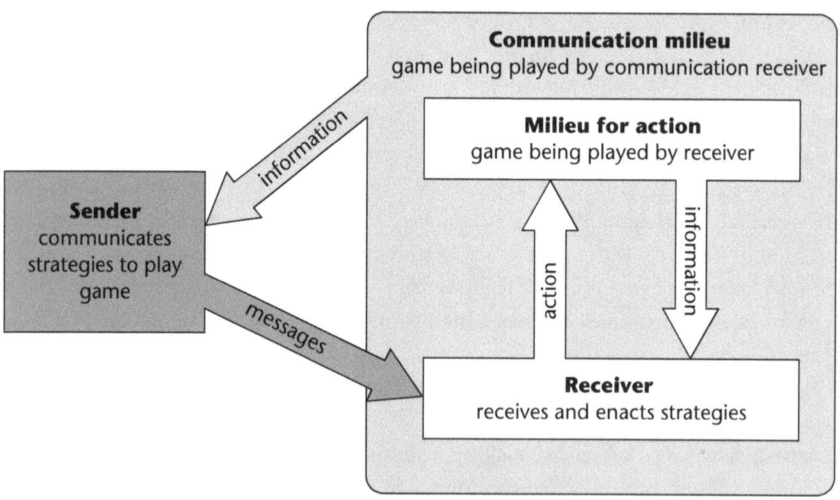

FIGURE 5.1 A representation of the situation of communication

An adaptation of Brousseau, 1997, p. 67

follows by saying the result of adding either 1 or 2 to the number said before; and so on, until one of them says 20, thus winning the game. The action task thus confronts an individual actor (learner) against a milieu of action composed, notably, of knowledge of addition, the rules of the game (e.g., one may not skip a turn or add 3, the goal is to say 20), and the play of his or her opponent (e.g., the opponent may play however they want as long as they abide by the rules). As Brousseau (1997) has noted, merely by playing the game, young students often discover that if one person gets to say 17, then that person can win the game. Yet at that time most of those "theorems" about the game and their related strategies for how to play are tacit, "theorems-in-action" (Vergnaud, 1996), embodied in the actual play but not necessarily spoken. In contrast with that a-didactical action task, a communication task takes place when the class forms two groups and designates representatives from each group to play the game at the board, with each round of play preceded by a phase in which each group confers, thus instructing their representative on how to play. In such a context, one gets a realization of the a-didactical communication task as follows: A communicator (learner) is a student in one of the groups who is not going to be at the board when the game is played; he or she may propose a strategy that supposedly would help the player win, such as "Always add a different number than your opponent; for example if she added 2, you add 1." The milieu of communication is composed of the group representative at the board (as receiver of the strategic message) playing the game against the representative of the other group. When that representative plays the game, she may enact the strategy proposed by the learner and through that enactment allow the milieu of communication to provide feedback on the validity of the learner's (communicator) strategy. (In this case, the communicator might realize that the strategy he or she communicated only helps win if the player has already said a number congruent with 2 mod 3, but that it would guarantee losing if the player had not yet said such a number). More generally, in an a-didactical communication task, the work of the learner consists of producing formulations (e.g., statements, diagrams) that are received by other students who decode them and enact them, with this enactment producing feedback on the quality of those formulations.

We use the notion of communication task to elaborate on how the study of geometry in secondary school might expand students' conceptions of figure. In our elaboration of examples we are less concerned with the adidacticity of the milieu (of action or of communication) and more concerned with showing simple communication tasks that summon existing conceptions and push them to evolve into more sophisticated ones along the lines of the orientation that secondary geometry provides the tools for modeling the experiential world. In particular, the general structure of communication tasks enables us to design tasks that include both a realm of action as reference (i.e., the child's experience with shape and space in the macro, meso, or micro space) and a realm of communication in which various kinds of signs (diagrams, language, symbols) are transacted in reference to the realm of action. The prior précis on the TDS serves to justify why we focus on communication tasks.

5.3. Secondary Geometry in the Service of Modeling the Experience with Shape and Space

In what follows we conceptualize a set of communication tasks that elaborate on the four conceptions of figure proposed in Chapter 2: Each of these communication tasks is meant to flesh out the notion that the study of geometry in secondary school consists of the development of mathematical models of geometric figures, where each of the four conceptions of figure provides a representation of the object being modeled. In the language of TDS, we could say that the action task we refer to is one that summons the conception targeted by posing a problem (or narrating a scenario which is in the sphere of practice of that conception), while the communication task that we propose is one that seeks the development of representations and controls with which students might model their experiences in that sphere of practice. Thus, we suggest that communication tasks can be used to design instructional interventions whose stakes include the development of knowledge that advances the project of modeling mathematically the conceptions of figure that students bring with them from earlier learning (and that were described in Chapter 2).

By modeling mathematically we mean two things. On the one hand, those communication tasks give students the opportunity to create models (in the form of representations that use diagrams, symbols, or language to depict what otherwise would just be enacted) of the actions of agents in the spheres of practice targeted by each of the existing conceptions of figure. On the other hand, those representations are subject to some reasoned or disciplined processing (e.g., logical argument, calculation) that informs the communication (e.g., produces a rational anticipation of experience).

We call these cases of student enactment of communication tasks *mathematical modeling* for two reasons. On the one hand, the tasks create conditions for students to move further in constructing a theoretical notion of geometric figure against the background of actions, tools, and concrete materials in space and shape with which they have been conversant before. In this sense, geometric figures as abstractions are models of concrete experiences with shape and space, and the work that students are to do in communication tasks is modeling in the sense of creating the abstractions that can represent those experiences. Obviously, we are not merely talking about a binary distinction of concrete vs. abstract but about a process of moving toward abstraction as models of geometric experiences (one that could pass through still concrete but more analytic stages, such as what Duval, 2005, calls dimensional deconstruction of diagrams). On the other hand, at each step of the way in the creation of models of the geometric experience, at whichever state of development of abstract concepts students are, they may be able to manipulate the representations at hand to make it possible to anticipate experience. In the extreme of complete formalization of the geometric experience, we might hope students model their experiences with shape and space through formal deduction of conclusions from symbolic representations and logical premises. In the extreme of no representation at all, that is in the extreme of only going through the experiences, young children may have no ways of producing a reasoned anticipation other than "I'll see when I finish". Yet at all the steps in between, where some representations exist, these representations will afford possible ways of manipulating them that are akin to reasoning (even if not purely deductive reasoning) and that permit those representations not only to be models but also to work as models, or to enable reasoned anticipations. We now elaborate what that modeling perspective could provide for the work of designing communication tasks that build on action tasks within background conceptions of figure.

5.3.1. Using Existing Conceptions of Figure to Represent Navigations of the Macrospace

The conception of figure that we called, in Chapter 2, figure as navigation of the macrospace engages students in action tasks that might include getting to a landmark in a city or comparing the length of the paths between two locations in a city. Action tasks that mobilize that conception may involve actual displacements.

A communication task might involve students in providing a set of directions orally or through a diagram, or in comparing trajectories by processing information provided in diagrams and instructions; in doing so the communication task might bring forth other existing conceptions of figure to model the macrospace and the navigation activity. For example, an oral communication might describe directions in terms of straight lines, number of blocks, and turns, all of which are verbal representations of figure; and a diagram might represent city blocks using rectangles and streets using rulers and straight lines.

One could then expect that some of the affordances of other conceptions (e.g., the conception we called "figure as construction in the microspace") might come in handy to represent such navigations (viz., knowledge of how to make a rectangle or how to draw a line becomes useful to create a city map or record a straight path on a map). The notion that two congruent segments measure the same length, which is an important element of the control structure of microspace conceptions, could be used to prescribe how to use the ruler to represent equidistant locations in a city map, and consequently to induce expectations in the navigation (e.g., two distances that are represented as equally long segments should take just as much effort or time to traverse) and the use of tools (e.g., compass, GIS) in the navigation. In other words, the diagram or the oral instructions become not only carriers of a message in a communication task, but also models, inasmuch as they bring with them the means of inducing more geometric properties into the representations available to the navigation conception. Consider as an example the following very simple scenario.

> Bobby went for a visit to his aunt and took his bike for a ride around the block. He was surprised to have to turn too many corners to reach his aunt's house. When he arrived he said he had thought he was lost. How could you explain to Bobby why he may have thought he was lost when he was not?

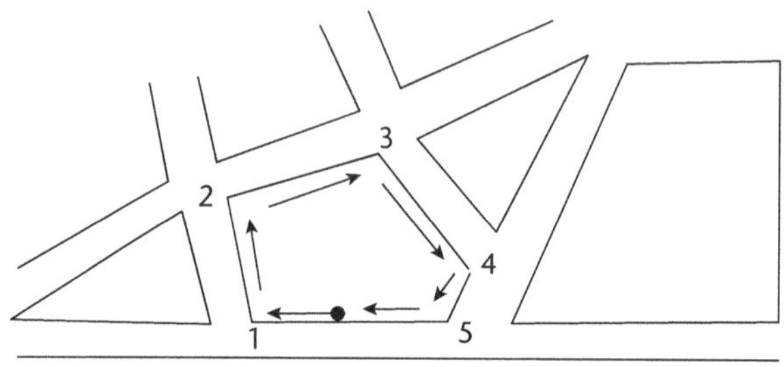

FIGURE 5.2 A model of Bobby's experience going around his aunt's block

We could expect secondary students to choose a square or rectangle to represent a city block. A triangle and other polygons become thinkable then, and with them the possibility to anticipate that the city layout around Bobby's aunt's house might include a pentagonal block. A representation of Bobby's navigation as traversing the perimeter of a pentagon or hexagon, which they draw with ruler and protractor, while counting the corners might suggest that the explanation for Bobby's experience is that he expected the block to be like a square.

To make a diagram on paper a student would use one of the microspace conceptions to model the navigation of the macrospace. While making diagrams is not the only way in which a communication task can proceed, the example above illustrates how communication tasks can summon other conceptions of figure to engage the learner in the mathematical modeling of a notion of figure. An important goal of secondary geometry instruction is to provide the means to understand and control the navigation of the macrospace; this includes the development of knowledge needed not only to make and read maps and architectural plans (which includes controls to wield the tools to make these diagrams), but also to understand and translate those messages in terms of driving or flying routes (see Jones et al., 2010; Kauffmann et al., 2000).

5.3.2. Using Existing Conceptions of Figure to Anticipate the Capture of Objects in the Mesospace

Microspace conceptions of figure can also be summoned to model the conception of figure as capture of objects in the mesospace. This conception involves students in managing spatial relationships with objects of size comparable to themselves. Consider the action tasks of moving furniture into a room or routing a cable from one location to another. The actor may need to decide whether a given piece of furniture might fit in a given space and allow for other actions (e.g., the opening of a door): These tasks might initially be done by trial and error. But students could also be involved in tasks of communicating how to do things that might involve them in describing or diagramming geometric figures. This in turn may serve to anticipate how the actions could be taken, or what actions should not be taken.

This work could get to more precise uses of the notions of angle (and tools used to make and measure angles) as illustrated by Fyhn (2008) in her description of how 7th graders used the notion of angle and diagrammatic representations of angle as they described their actions when engaged in rock climbing. It could also get to more precise and accurate descriptions or diagrams of objects in their spaces by employing the operations and systems of representations from microspace conceptions into these communication tasks. The task described by Berthelot and Salin (1998), which took students to collect information about a bench and then locate where that bench might be placed when moved, gives an example of what these communication tasks could look like. An important goal for secondary geometry instruction is to develop in students the capacity to analyze large objects that might initially be seen as single things (e.g., a table, an exercise machine) as composed of smaller primitives that can be described using basic geometric elements and hence can be controlled, in the sense of understanding how they might behave when transformed (rotated, translated) in space. This capacity to see the real in terms of the geometric seems essential for contemporary jobs such as graphic and industrial design (let alone to understand geometric theory, as explained by Duval, 2005). In turn, that analytic reduction of large objects into geometric primitives could help summon representations and controls from the microspace conceptions in order to anticipate the outcomes of action.

The existence of systems that allow the modeling and manipulation of objects in virtual 3D environments (e.g., Moustakas et al., 2006; Spikol & Eliasson, 2010) permits us to think about interaction with and communication about a multitude of 3D objects, where the language of description associated with the conception *figure as description of objects in the microspace* can be brought over to induce a geometry into the objects of the mesospace. This language of description can be visual as well, and eventually combine with the operators from the construction conception, as when 3D primitives (e.g., sphere, cube, cone) are used as the basis for describing a complex 3D object, and design software is later used to construct a model that looks like the original. Laborde (2008) has shown how software especially designed for students to learn three-dimensional geometry can support the analysis of three-dimensional objects in terms of lower dimensional geometric concepts (see Duval, 2005).

5.3.3. Improving the Control of Microspace Conceptions

The modeling perspective can also guide the development of the two conceptions of figure from the microspace: *figure as description and movement of small objects* and *figure as construction of small objects*. We have seen how those conceptions could provide the semiotic systems for communication tasks in the macro and mesospace, partly because those conceptions have already been operational, earlier on in elementary school, in activities that involve some amount of communication. At the secondary level we would expect the systems of representation of those conceptions to become more sophisticated. Take for example the conception of figure as description and manipulation of small objects. Early on, this conception can be seen at play in action tasks such as those of a young child fitting shapes into the openings of a shape sorter (Fisher et al., 2013; Verdine et al., 2014). Children's tactile and visual coordination may help them succeed in shape-sorting. They can then use signs (e.g., deictic gestures such as pointing, and deictic expressions such as "that one") to request desired shapes. As the van Hiele-based research has shown (see Chapter 3), given common experiences in school, students recognize shapes as wholes before they can describe their properties. In elementary school, students acquire language to describe the shapes they have been playing with since childhood (e.g., names for whole figures, such as square, and for some properties, like parallel); they also learn to measure lengths and angles and to describe figures in terms of their measures.

As regards the conception of figure as description and manipulation of small objects, secondary geometry instruction has to develop students' language of description further, so as to enable students to describe those shapes and new ones not only in terms of their names and dimensions but also in terms of the relationships among their components (e.g., synthetic properties) and in terms of relationships with external references (e.g., coordinates of location). In particular, a goal of secondary geometry instruction is for students to learn the definitions and properties of 2D and 3D geometric shapes, and to be able to use them to discover and prove entailments of the properties of shapes as well as set relationships between shapes (e.g., inclusion, disjointness). Communication tasks in which students need to process (reason with, calculate with) descriptions of an object in order to find out new things about those objects serve to create analytic models of those shapes, particularly moving students up in a learning trajectory.

Likewise, the conception that we have called *construction of small objects in the microspace* starts early on in childhood with drawing and play with construction sets (e.g., Lego or wood blocks) and is later enriched in elementary school mathematics and art with learning to use instruments such as straightedge and compass, stencils, drawing software, and so on. As they grow, children enrich the set of categories they pay attention to in their drawings (Karmiloff-Smith, 1990) and as they gain occupational dexterity they can manage to create more detailed and precise objects, including diagrams and 3D replicas, turtle graphic drawings, and digital drawings.

As regards the conception of figure as construction of small objects, in secondary schools one could expect students not only to master construction procedures and explore generality through dynamic geometry constructions, but also to develop reflexive and critical capacity on their means to achieve those constructions, for example by being able to identify and visualize separate features of constructions, anticipate entailments and contradictions of steps in constructions and so on.

The system of representation of this conception (strokes, dots, etc., in the case of drawings) can serve as a semiotic system for students, with which they can author diagrammatic propositions that model figure as navigation of the macrospace and figure as capture of objects in the mesospace. Yet, at the secondary level, students can also expand and better organize this semiotic system by developing the means to control how the use of those signs participates in making meaning. Specifically, communication tasks that prescribe or describe a construction can help students incorporate notions of stipulation, constraint, and necessity to the way they control their construction activities and provide them with ways to manage the demands of discipline and accuracy. Indeed, the "pragmatic reasoning schemas" proposed by Cheng et al. (1986) as apt scaffolds for the learning of deductive reasoning can find quite congenial embodiment in the work of construction (see also Stylianides and Stylianides, 2008).

5.4. Communication Tasks in the Teaching and Learning of Geometry

In this section we elaborate on the ideas introduced in the prior section, providing examples that might be used by Masters students or practicing teachers in research-in-practice projects. We have in mind design research efforts that might be carried out iteratively in a single classroom and that might lead to better understanding of the affordances and challenges of communication tasks as well as illustrations of what types of learning and performance might be observed from students. An understanding of design research would be needed to carry out these projects but it is beyond the scope of this book to provide that. Interested readers could consult McKenney and Reeves (2012) or Van den Akker et al. (2006).

The tasks suggested vary in the extent to which they have already been researched; some have been tried, and documentation of their trials is cited to inform future work, while others are only proposed here with some suggestions of why they might be of interest. For those reasons the prose is not uniform: In some cases we report more than argue, in others we do the reverse. The suggestions have more elaboration in task design and learning goals than in research instrumentation or analytic perspectives. We expect that readers who become interested in undertaking a design research project related to one of these tasks will work closely with their advisors to make those elements more precise.

The four conceptions of figure identified in Chapter 2 enable us to think about the following generic action tasks

1. To move about a space
2. To capture (e.g., visually) a large object
3. To describe features of a small object
4. To construct a small object.

Those action tasks can be connected to communication tasks, utilizing systems of representation from other conceptions. Generically that might mean to create a message that enables the receiver to engage in one of those action tasks. In particular we could unpack those as

1.1. To describe or prescribe a navigation of a space
2.1. To describe or prescribe the capture of a large object

3.1. To create or discover a code or language for the description of small objects
4.1. To identify the controls needed to describe an accomplished construction or prescribe a construction procedure.

In what follows we try and make those communication tasks more specific by proposing examples and drawing connections to prior chapters and to the literature. It should be apparent here that, inasmuch as those communication tasks involve acts of description or prescription, they require the mediation of a system of representation that could be verbal, but that could also include diagrams, gestures, and concrete models. While our use of "representation" is broad, to include many ways of representing, in all cases it relies on material artifacts (even if embodied) as representamens or signifiers (Peirce, 1897/1955).

Obviously, we are not covering all sorts of tasks that could be investigated or implemented in secondary school, but limiting ourselves to communication tasks. In particular, the informed reader will note that in regard to the construction conception we are not addressing nearly enough of the extensive literature on dynamic geometry tasks, though some of this literature has been reviewed in prior chapters. Readers can consult Laborde et al. (2006) or Sinclair and Robutti (2013) to complement their knowledge of that specific area of research.

We present below sketches of possible activities that might allow each of the four conceptions to evolve through modeling in the senses described above—creating representations of figure and manipulating those representations in the service of rational anticipation. In each case, we spell out not a complete intervention but rather a sketch of activities and questions that connect basic conceptions of figure through communication tasks and, hence, promise to perturb those basic conceptions in some way that might allow for growth. Clearly these are not assurances but rather conjectures, and we feel free to present them here because this is a textbook that might inspire young scholars (or adventurous teacher-scholars) to investigate the unknown. The sketches of tasks allow one to ask questions that we think are fun and relevant, but they provide no assurances of effectiveness. These activities are not organized in a sequence that suggests how they could be chronologically laid along the curriculum, but rather described in an order that makes sense from the perspective of the four conceptions listed in Chapter 2 and their evolution anticipated earlier in this chapter. Our goal at this stage is merely to articulate what isolated interventions might look like that would challenge students' existing conceptions of figure, not yet to develop them fully and organize them in a chronology that makes sense from a curricular perspective. Those larger design tasks could benefit from design research emerging from the suggestions hereafter and could eventually support the development of curricula.

5.4.1. Description or Prescription of a Navigation of a Space

While the actor in the navigation of the macrospace enacts a self-centered geometry, a communication task can promote taking some perspective from such enactments, representing those enactments as diagrams, and using those diagrams to make inferences about enactments. The task proposed here aims at promoting the use of microspace drawings to model simultaneous macrospace navigations, and to use the affordances of the model to make inferences about possibilities in the macrospace.

> Consider three English-speaking friends (Huey, Louis, and Dewey) who are left in the very center of Plaza 25 de Mayo, the main square in the city of Resistencia[2] (the capital of the Chaco province in Argentina). The map below is provided for the reader to envision what the task might look like, though students might be given less information (e.g., they could just be told that the main square is the size of four city blocks and that four avenues—9 de Julio, Sarmiento, 25 de Mayo, and Alberdi—spring from the midpoints of the sides of the square).

Students are told that the three friends separate and walk around the city. When they come back together they narrate what they did. The following accounts are initial drafts of what those narrations could be; they are merely illustrative, and

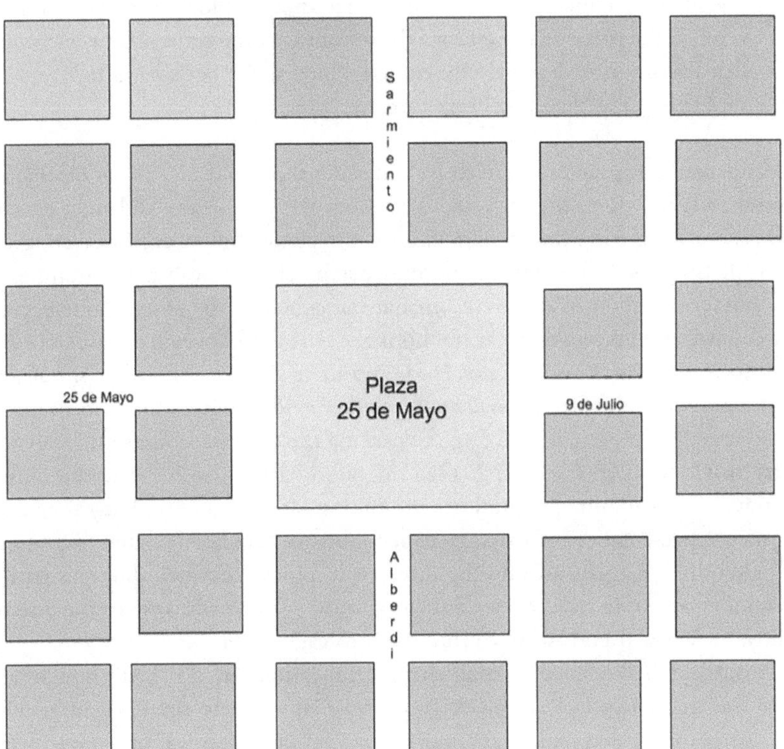

FIGURE 5.3 The layout of the civic center of Resistencia (Chaco)

teachers should craft them carefully as they are the givens for the problem assigned to secondary students.

> Huey says he walked southeast to the end of the plaza, walked two blocks along Av. 9 de Julio, turned right and walked five blocks, turned right again and walked five blocks (somewhere there he entered a bookstore and stayed there for 10 minutes). Coming out of the bookstore he turned to his right again for two blocks, and to the right again for one block, then turned left and walked three blocks to reach the plaza.
>
> Louis says he walked southwest to the end of the plaza, kept going on Av. Alberdi for three blocks, turned left two blocks, then right one block, and right again for five blocks. Somewhere there he entered a bookstore and waved to Huey, but Huey did not see him. Louis stayed at the bookstore for 10 minutes and left, turned right again for eight blocks, made a right and walked three more blocks, and then right again, walking to reach the plaza.
>
> Dewey says he walked north to the end of the plaza finding himself in a corner, took the street to his front right (going northeast) for three blocks, turned left for four blocks, then left for five blocks, and left again for six blocks, then left one block. He stopped for coffee for five minutes. Upon finishing his coffee he walked to his left another block, then right until he reached the plaza. Having heard the accounts of the others, Dewey contends that Louis mistook somebody else for Huey at the bookstore; he's not even sure that it was the same bookstore.

Students could be given subjective accounts like those and asked a set of objective questions: If the three friends walked at about the same pace, did they ever run into each other during their walk? What is the closest they were to meeting each other? If they took about three minutes per block, for how many minutes did they miss each other? Who arrived first at the center of the plaza? Is Dewey correct about the misrecognition at the bookstore? Is there a chance it was the same bookstore? The questions are not hard, and their geometry is rather simple. But they require the solver to rationally put together evidence from those accounts into a new system of representation (e.g., translating the verbal accounts into polygonal paths to which, as in Figure 5.2, they might add arrows to indicate the sense of the traversals or numbers to indicate the timings at each point). They then could use such representation of space-time accounts to calculate possible responses to the questions—e.g., measuring the distance in blocks between different paths at the same moment in time, then comparing those numbers to answer the question of how close the friends were to running into each other.

The research literature in mathematics education has not had enough studies of the teaching and learning of geometry that relate to the navigation of the macrospace. Clements and Sarama (2014, p. 124) provide a learning trajectory for spatial thinking that includes children up to eight years of age and that suggests children's capacity for map reading and map making (see also Kotsopoulos et al.,

2015). The literature on computer-assisted design, visualization, and geography contains important starting points for tasks and resources that could be used in interventions with students (Liben, 2007; Liben et al., 2002; Uttal et al., 2006), yet there are important mathematics education questions to ask about such interventions, to which we come later in this chapter.

An important question, connected to what was said above regarding instructional situations, is how the task might be anchored in the usual situations of a geometry class. It is apparent, for example, that the geometric face validity of the task itself is low, and a teacher might feel the need to justify the task as part of the geometry course. The teacher could do that by providing the city map, which in this case helps connect the task posed to a situation of exploration. Alternatively the research could explore whether changing the context of a navigation in the city into one of people navigating an open terrain using compass directions might allow for similar reasoning and could be anchored in a situation of construction (where a way of scaffolding the task could include telling students to represent or sketch the scenario). Students might use a number of different resources to create their representation, including gestures, geometric language, 2D or 3D diagramming software, DGS, or construction materials such as Lego or wood blocks. Each of these options, in turn, could support their development of skills in tool-use for the conception of figure as construction of small objects, which is being used to model the experiential world.

5.4.2. To Describe or Prescribe the Capture of an Object

In an action task that involves the conception of figure as capture of an object in the mesospace, the actor might have to measure, seize, manipulate, or navigate around a mesospace object. The task offered by Berthelot and Salin (1998) and described in Chapter 2, in which students have to mark the locations where the feet of a bench will go when moved before actually moving it, is an illustration of what a communication task might look like. Gonzato et al. (2011) have identified other tasks (see also Gonzato and Godino, 2010). Here we provide other possibilities.

A communication task to mobilize this conception of figure could involve a student (producer) in creating a description of the way another agent captures an object or prescribing how another agent (human, mechanical) might handle the object. By capturing we mean both perceiving (visually, haptically) and manipulating (displacing) a physical object as a whole. Consider the case of finding out something about a large object (e.g., a stack of boxes) from reading a message about that object (e.g., looking at a photograph or sketch of that object). Consider that there is such a stack of boxes somewhere and that the students are asked to produce a diagram (e.g., they might eventually produce a perspective diagram like the one provided in Figure 5.4) for a receiver to determine how many boxes there are. Alternatively the producer might have to produce a set of messages to describe to a forklift robot how a set of boxes needs to be stacked.

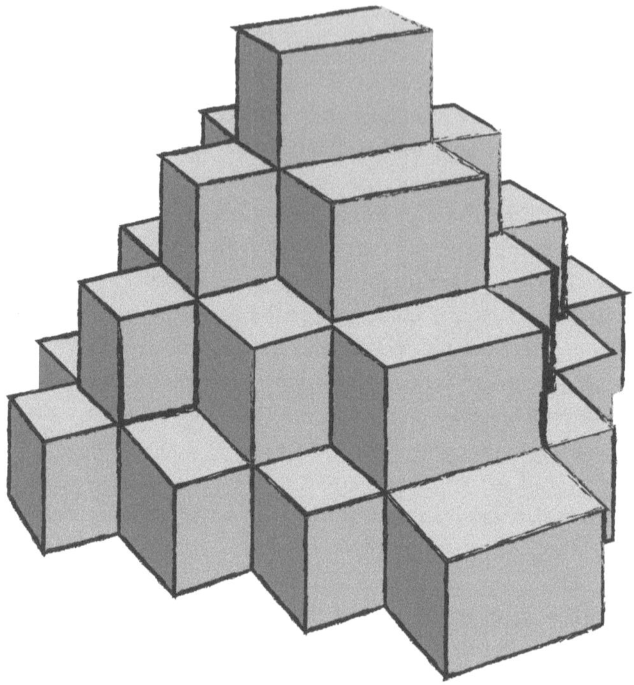

FIGURE 5.4 A stack of boxes

As in the case of messages about navigations, several systems of representation may be used to describe or prescribe how to handle large objects, including words, diagrams, 3D artifacts, and virtual assets. In describing or representing how to handle such large objects, students may bring in their knowledge of geometric figures in the microspace to inform the way they refer to large objects. In particular they may use the descriptive language to switch from a synthetic view of the whole object to an analytic view of parts of the object, each of which might be described (viz. represented) using elements of geometric knowledge, what Duval (2005) calls dimensional deconstruction.

Take for example the problem of producing a diagram such as the one above, indicating how a set of boxes has been stacked: This could involve the student in analyzing their knowledge of what a box is using their knowledge of quadrilaterals to create a perspective drawing of a box—using, for example, their knowledge of how to draw parallel lines to represent parallel edges in boxes. Alternatively, the student could be asked to describe this way of stacking the boxes so that a receiver of the message could produce an equivalent stacking. This could give context to the producer to develop or reuse an analytic way of describing each box, for example, in terms of vertices, edges, and faces, as well as opposite faces, opposite vertices, concurrent edges and so on. This activity could help students to transition to more sophisticated versions of this conception.

In what way could these communication tasks pave the way to engaging students in the kind of mathematizing that is usually expected in secondary geometry, that is, in defining, conjecturing, and proving? The distinction between *knowing* and *seeing* explained in Chapter 2 (Parzysz, 1988) can be useful as a way to conceive how communication tasks might be used to engage students in not just recognizing and applying geometric knowledge but also in making arguments about it. The drawing used above to represent the box stacking message instantiates how 3D geometric diagrams can facilitate "seeing" while also eliciting some errors students

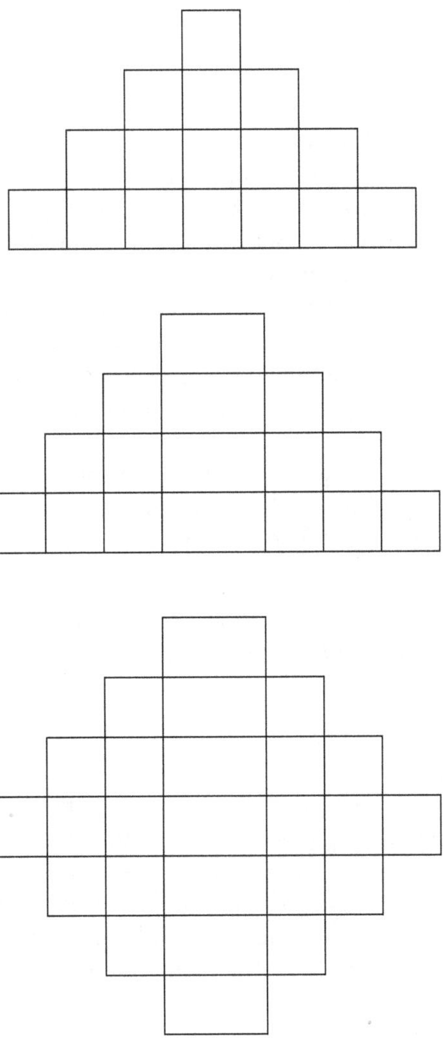

FIGURE 5.5 Side, front, and top views of a stack of boxes

make (see Chapter 3). A completely different message could be provided by a set of side, front, and floor diagrams that also represent (the same) stack of boxes, such as those given in Figure 5.5. Students could be asked to count, build, and (possibly) also stack actual boxes that respond to the dimensions specified in those diagrams. While the diagrams are accurate as two-dimensional diagrams and permit counting and measuring shapes, none of those shapes by themselves represent (in a "seeing" sense) the objects to be counted, constructed, or measured. (For example, while one can see 16 squares in the side view, there are not just 16 boxes, and the faces of the boxes are not all squares). Instead, those images and their counts and measures need to be put together into a different representation so that a claim about the geometric objects can be justified rationally: This new representation would put together information that may have its basis in taking on some empirical (visual) evidence and ignoring other such evidence.

The example illustrates how these communication tasks can be further pushed to promote the kind of rational control of experience that we identified earlier as characteristic of the modeling approach we are proposing. It hinges on developing ways of organizing and questioning the information provided by the diagram (or the message, more generally), for example by creating experiences that require reconciling diverse accounts or using information from diverse accounts to produce new knowledge. The extent to which some information is empirical in origin (e.g., the student might measure or count something in a diagram) and the extent to which some objects may not be completely formalized (e.g., a box is still not constructed as a mathematical object and, in particular, some assumptions about boxes, such as that their parallel faces are congruent, are still tacit) are less relevant. Yet, the work of putting together the information known in the service of answering a question about the boxes is a piece of genuine geometric modeling work and involves reasoning. This work, paraphrasing Netz's (1998a; see Chapter 2) comment about the Euclidean work with diagrams, is rational even if not logical.

5.4.3. To Create or Discover a Code or Language for the Description of Geometric Diagrams and Small Objects

By the time students come to secondary school they already have the basis for a language of description: They know names for shapes such as "square" and "sphere" and they know diagrams and other representations of those shapes. In our attention to the prior two conceptions, we described how this language of description could be used as semiotic resources in communication tasks like the ones described above, to actually make more analytic the conceptions of figure as navigation of the macrospace and figure as handling of objects in the mesospace. In this section we deal with how communication tasks in secondary geometry can also help evolve the conception of figure as description of small objects in the microspace (an evolution that may start earlier on, as Sinclair and Moss, 2012, describe).

While the evolution of the earlier two conceptions is described in terms of making their representations more analytic,[3] so as to permit the solving of communication problems, here we are concerned with the system of representation as a whole. How can communication tasks about descriptions of figure help further develop the language of description of figures in the microspace?

We contend that communication tasks can prompt working out the complementarity and coordination of descriptions: For example they can prompt the need to coordinate the relationship among words like *stroke*, *segment*, *line*, *side*, and *edge*, which call for the establishment of definitions and the statement of properties. They can also prompt such coordination between different ways of describing figures, including diagrams and algebraic expressions (as in geometric calculations in algebra; see Boileau and Herbst, 2015). We consider several examples below.

5.4.3.1. From Names and Prototypes to Properties

One way in which a language of description needs to evolve, somehow foreshadowed by the transition from level 1 to level 2 in van Hiele levels, is from holistic representations of the figure (through names and prototypical diagrams) to representations of figures as sets of concurrent properties. How could this transition be prompted?

Consider the game "Guess my Quadrilateral" (an adaptation of the game "Guess Who?"), reported by Herbst et al. (2005). In this game, one person takes a picture of a shape out of an envelope (the shape is a quadrilateral, given by a prototypical diagram and its name) and other students (who cannot see the shape) need to guess what shape it is. Toward this end, guessers can ask questions that admit the answers Yes or No. They could directly ask the question "Is it a [name of quadrilateral]?" but they accumulate many penalty points if they miss; thus there is an incentive not to ask that question until they are ready to guess with a high degree of certainty of being correct. Instead students tend to ask questions that do not name the quadrilateral directly. For example, they ask: Does it have parallel sides? Does it have a right angle? Students accumulate points depending on the number of questions they use until they find out what the shape is; and the student with the least number of points (that is, whoever guesses using the least number of questions) wins. In the context of this game, ninth graders (14 years old) have been observed to analyze quadrilaterals in terms of their characteristic properties, not only asking relevant questions but also optimizing those questions so that they produce maximal information in the smallest number of questions. Properties used include congruence and parallelism of sides, congruence, perpendicularity, and bisection of diagonals, congruence of angles, and axial symmetry. This simple game illustrates how the conception of figure as description of small objects can transition from description as names and prototypical diagrams of figures to description as set of concurrent properties of figures.

That simple case also illustrates a general point about the communication task: It can spur work in which a language of description is generated and further developed. By *language of description* we mean not only the development of lexical elements for shapes or their prototypical diagrams, some of which students are likely to bring with them from elementary school. We rather mean the development of lexical elements for the properties of those shapes and of a grammar that ties those elements beyond clauses like 'this is that' (e.g., "this is a circle" or "*O* is a circle") used to name shapes at younger ages. We expect that students come to secondary school with such a lexicon and basic grammar, yet that their specialized grammar will grow more sophisticated in secondary geometry.

In secondary school we can expect further development of that lexicon and grammar. Indeed Duval (2005) notes that "the cognitive articulation between the register of visualization and that of language is not done at the level of words but at the level of propositions" (p. 33, our translation).[4] How do these communication tasks engage the student in further expanding the register for modeling their experiences with shape and space? In the "Guess my Quadrilateral" game two representations are "known" to the person that has the shape: The diagram and the name. (The person who is guessing the shape does not know which particular shape they have to determine, but he or she knows the whole set of them, both in terms of prototypical shapes and in terms of names.) But the strictures of the task, the requirement to ask questions that can be answered yes or no, push students to model what each of them (diagram, name) means in the form of a set of concurrent properties. These need a much more complicated grammar than that of designation.

One could take the game as a prototype of the modeling work: The "reality" being modeled is a less sophisticated representation system including a set of diagrams and their names, while the modeling language consists of the names for elements of the diagrams (e.g., side, angle) and questions or statements about the possible properties those elements might have. In principle, these properties could be spatiographical (e.g., "is one diagonal 10 cm. long?") as well as theoretical (e.g., "are the diagonals congruent?"), and the way of ascertaining those properties could be empirical (e.g., to check that two diagonals are congruent the student might measure both) as well as deductive (e.g., to verify that all sides are congruent the student could reason that if opposite sides are congruent and consecutive sides are congruent, all sides are congruent). Further if two sets of properties identify the same figure, one could ask whether the sets are equivalent, or whether only one of them allows to deduce the other. The "Guess my Quadrilateral" game thus illustrates a communication task in which descriptions of a geometric object can be questioned as to their meaning (e.g., when somebody says "yes" to a question that includes a disjunction, what does that mean?), or enter competition with each other (e.g., given two statements about the same issue, which one provides more information?). This work might aid in developing the description conception from a state in which objects are recognized by names and shapes to a state in which objects are recognized as sets of properties and even properties can be "seen as" (Coliva, 2012) different properties (e.g., two concurrent sides may also be seen as part of two intersecting lines). To make these sets of concurrent properties do their work, ordinary language needs to be harnessed and disciplined.

5.4.3.2. From Seeing Diagrams to Stating Properties

Beyond developing an analytic language of description, secondary geometry needs to induct students into the statement and proving of properties about figures. The "Intersecting lines" lesson studied by Chen and Herbst (2013; see Figure 5.6) provides an example of how a communication task could get this work going. In the "Intersecting lines" lesson, ninth grade (14-year-old) students were given a sheet of paper with six lines drawn (see a reproduction below), of which some intersect on the paper and others do not. Some of the latter appear as if they would intersect if extended, while others appear to be parallel, yet nothing is explicitly taken as known about any of them being or not being parallel. Students are asked to determine how many angles they would have to measure in order to know all the angles made when each of the lines crosses each other. As can be noted, the task used by Chen and Herbst (2013) does not quite seem like a communication task. While it exposed students to some interesting ideas and reasoning, we would suggest that a second iteration could frame it more as a communication task.

In the context of this task, students have been observed to act as if they were holding as true (this is, what Vergnaud, 1996, would describes as holding a *theorem-in-action*, or what in the language of conceptions can be described as using controls) properties like: (1) if one knows the measure of an angle, the angle vertical to that angle can also be known without measuring, and it has the same measure; (2) if one knows the measure of an angle, the measure of the angle adjacent to the known angle can also be known without measuring, and its measure is 180 degrees minus the measure of the known angle; (3) if three lines intersect pairwise and one knows

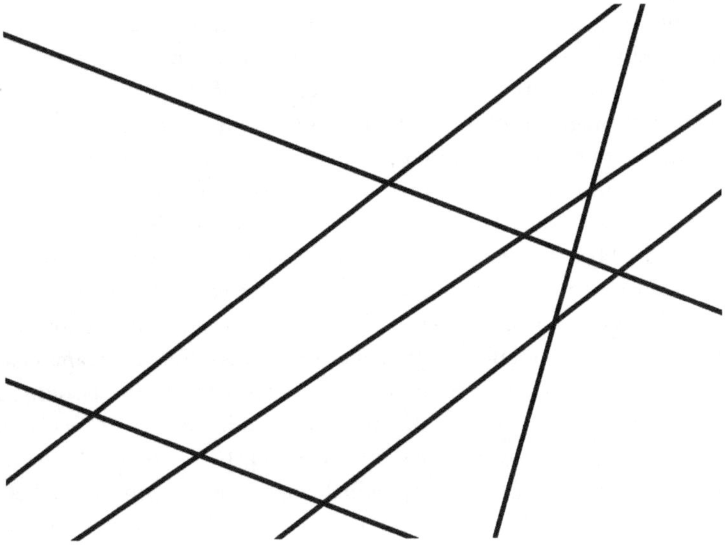

FIGURE 5.6 Six intersecting lines

(drawn according to the diagram used by Chen and Herbst, 2013) the angles at two of those intersections, the third angle can also be known without measuring, by subtracting the two known measures from 180 degrees. While those theorems-in-action allowed the students observed by Chen and Herbst (2013) to determine many angles, they were not enough for all angles. Particularly pressing are angles that appear to be congruent but are contained in different intersections (e.g., corresponding angles) and angles that are invisible (e.g., if lines intersect beyond the boundaries of the printed page). Students were observed also engaging in further reasoning that underscores their use of deduction from assumptions to model the set of angle measures to be found: One could describe the work of the student as modeling in the sense that they are using statements about expected relationships between measures of angles to predict what the actual measures of those would be.

Students have been seen hypothesizing that two lines would meet outside the page and finding a third line that meets those two inside the page, and in that context they have used the property of the sum of the angles of a triangle (a property they have known since middle school) to determine what angle the lines make if and when they intersect outside the page. So, if the angles the two lines make with the third line add to less than 180 degrees, they claim that the angle at the invisible intersection measures the difference, and if those angles add to 180 degrees they claim that the lines are actually parallel. Thus, this modeling work takes the actual measures of angles formed by lines on the page as the world to be modeled, and the properties and relationships between angle measures as the modeling language. Along with deduction and some initial information (e.g., the actual measure of a handful of angles), those properties and relationships enable students to find out (predict) what the measures of all other angles should be.

The "Intersecting lines" task was not robust enough to ensure the production of such reasoning. Hamlin (2005) showed how in one informal geometry class, the task was reduced to a very basic case of the situation of exploration, in which students were encouraged to measure. Thus, we recommend investigating variations of the task that cast it more like a communication task. Consider, for example, separating the person who has to find all the measures from the person who can (use tools to) measure the angles, as follows. The class is divided into several teams, each having six lines disposed in similar ways as shown in Figure 5.6, though not exactly equal to each other; there is one student, the measurer, who is not in any team and who is the only one close to all the diagrams and who can use a protractor to measure angles as requested by each team. In this setting the teams take turns asking the measurer to measure an angle, and then can use the resulting measurement on their own to deduce the measures of other angles. It seems that the constraints of this game would make it more desirable to (1) deduce those measures that can be deduced and (2) strategize which angles might produce the best yield if their measures were known. The game could end when one team has finished determining all their angles. As might be expected this task is also more novel than the original Intersecting Lines task, and it would be interesting to attend to how it taxes the work of teacher and students.

5.4.3.3. From Dynamic Explorations of Diagrams to Properties of Figures

As a third and last example, we consider the uses of dynamic diagrams, as they could be realized in DGS (dynamic geometry software; see our review in Chapter 3) or in videos that record transformations of diagrams or models over time (see for example Vi Hart's Hexaflexagons at https://www.youtube.com/watch?v=VIVIegSt81k). The general point is that a dynamic exploration of a figure in the form of the dragging of a dynamic diagram or the videotaping of the transformation of an artifact over time can serve to represent properties of a figure related to (e.g., represented by) such a diagram or model. To be more concrete, consider the following communication task: A student receives a dynamic sketch and the statement of a property (for example, they receive the sketch shown in Figure 5.7 and the statement *A quadrilateral inscribed in a circle has supplementary opposite angles*) and are asked to create a screen capture of their exploration of the diagram so as to communicate the property to a friend without using language. Alternatively they could just show the exploration without saying or writing anything.

Note that while this communication task uses a construction made on dynamic geometry and the work of the producer of the communication is likely to include dragging, our focus is on dragging as an element of the language of description, specifically dragging as a semiotic resource in the service of warranting equivalence

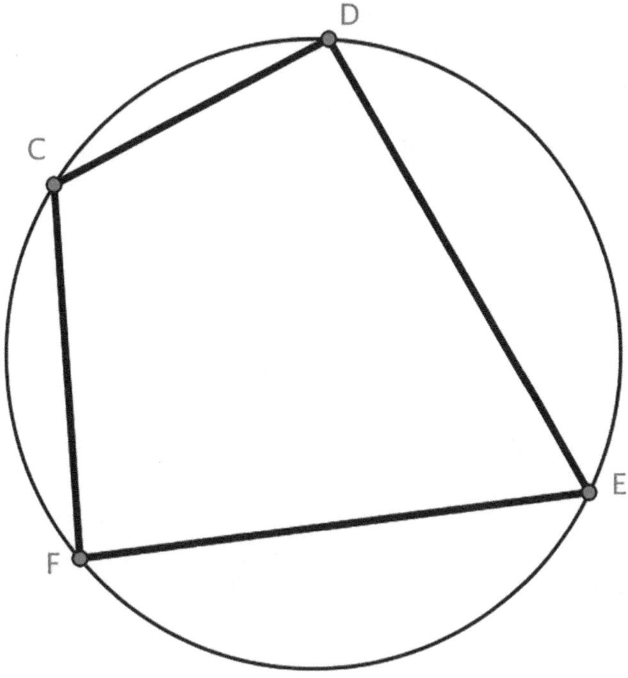

FIGURE 5.7 Givens for a communication task using DGS

in spite of difference (Battista, 2008), much in the same way that synonymy does it in language. For those reasons we include this use of DGS here as opposed to in the next section.

Leung (2003; also reviewed in Chapter 3) describes this use of dynamic sketches in the general terms of variation theory (see Runesson, 2006), noting how the dragging that tracks a desired property (what Arzarello et al., 1998b, called "lieu muet dragging") can serve the representation of a property in terms of synchronic simultaneity (i.e., concomitance of properties in space) and diachronic simultaneity (i.e., fulfillment of a property over time). Our example is a case of the former: Imagine our producer has set up the sketch to exhibit the measure of the angles and recorded the dragging of point *D*, obtaining as part of the screencast the diagrams shown in Figure 5.8.

The viewer presumably sees the change in the four measures as *D* is dragged on the same circle (maintaining the radius constant). Variation theory would identify this demonstration as a representation of the property at stake as a synchronic simultaneity, albeit one that is still somewhat opaque, in that the receiver needs to have the idea of adding opposite angles. Different objects in space—the pairs of angles, a pair of constant angles (at *D* and *F*) and a pair of variable angles (at *C* and *E*)—are simultaneously equal to straight angles at the same time that they are all on the circumference. (Gonzalez & Herbst, 2009, discuss other examples that could be classified under these same considerations).

In contrast, the example Leung (2003) discusses, the set of all circles that contain two given points (see Figure 5.9 below), illustrates a different way in which the lieu muet dragging of a dynamic diagram could communicate a property, what Leung (2003) calls *diachronic simultaneity*. As the center point *C* is dragged to represent all the circles that pass through points *A* and *B*, the viewer gets the opportunity to grasp the property that the locus of *C* is the perpendicular bisector of $\overline{AB}$. Every moment in time contributes a different bit to this representation of

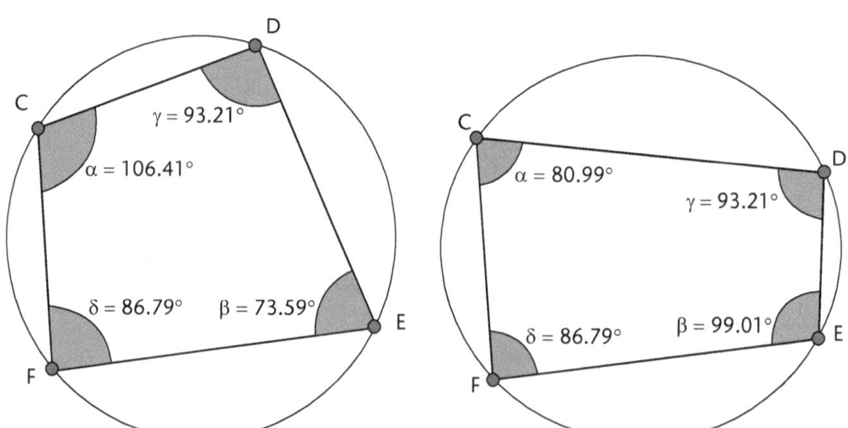

FIGURE 5.8 Angle measures for quadrilateral *CDEF* when dragging vertex *D*

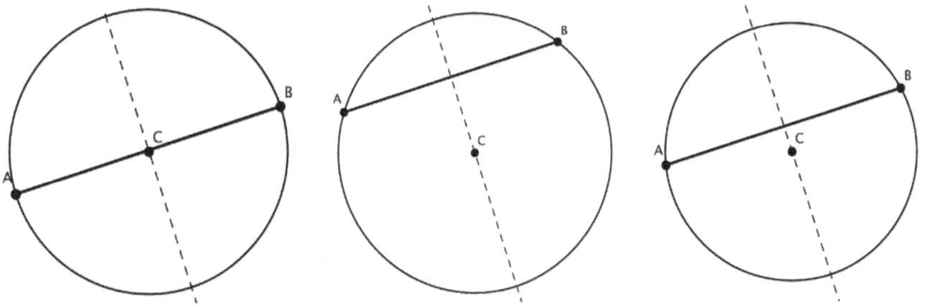

FIGURE 5.9 Three of all the circles that contain points A and B

the perpendicular bisector (C traces it over time) while at each moment in time there are simultaneities to observe, namely that the circle passes through the two points, that the segment is the same, that C is the center of the circle, etc.

The main point here is that the language of description of geometric diagrams can now be extended to include dragging (and screen casting) or manipulating and video recording as a resource to exhibit invariances against the background of difference.

5.4.3.4. Summarizing the Three Cases

Thus the work on creating a language of description for geometric objects can be facilitated with communication tasks that enable students to look at diagrams or geometric objects in two different ways (the modeled world and the modeling system). In the case of the "Guess my Quadrilateral" game, those two ways include, on the one hand, figures as whole shapes (represented by their names and prototypical diagrams) and, on the other hand, figures as sets of statements of concurrent properties about the elements of those shapes. In the "Intersecting lines" lesson, those two ways include, on the one hand, figure properties as dependent on empirical measures and, on the other hand, figure properties as also dependent on necessitated measures, predicted with the assistance of assumptions, initial conditions (possibly empirically determined), and, notably, deduction. Finally, in the case of dragging, those two ways include figure as a collection of diagrams related to each other over time and figure as a set of relations among elements of a diagram at each moment in time (these two ways combine in two different ways for the two cases of synchronic and diachronic simultaneity). In all of these cases, the work of reconciling both ways of describing the figure can not only help students move to more nuanced and operational knowledge of geometric figures (modeling in the sense of creating representations), but can also help them use informal proof as a tool to know with, even if some of the information that is processed in those informal proofs is, as in the cases of navigation of the macrospace and the handling of large objects in the mesospace, initially derived from experience (modeling in the sense of anticipating properties). From a cognitive perspective, these activities can help promote students to higher levels in a learning trajectory.

5.4.4. Identifying the Controls Needed to Describe an Accomplished Construction or Prescribe a Construction Procedure

By the time students come to secondary schools, they have experienced construction tools (such as the ruler and the protractor, the geoboard, building blocks, sticks and marshmallows, turtle graphics, etc.) and possibly even some construction procedures in art and mathematics classes. However, their means of controlling their constructions, at least in North America, are likely not to be very sophisticated. In secondary schools students learn to use more sophisticated tools such as the compass and DGS, and these by themselves may help the construction conception evolve, at the very least in terms of extending the set of operators and the system of representation. Along the lines of the modeling perspective we are articulating here, though, we contend that a main goal for communication tasks is to engage students critically and reflectively in the work of construction—in eliciting, exercising, and developing their means to control the constructions they enact.

As Schoenfeld (1988) noted, even successful, older students continue to build knowledge of constructions as the execution of procedures, as if this knowledge had nothing to do with the declarative knowledge from theorems. As a result, students might not have ways of calling out constructions that look accurate but are mathematically incorrect, or valuing constructions that look messy but are mathematically correct (see also Tam & Chen, 2012). We contend that secondary geometry presents the opportunity to improve, in particular, the control structures of the conception of figure as construction of small objects in the microspace. In the following subsections we sketch out what some of these controls could be and sketch out communication tasks where those controls could be developed. The following aspects of construction work might be described as a way of unpacking what we mean by engaging critically and reflectively in the work of construction.

Clearly, quite a bit has been written about geometric constructions in the last 25 years, especially owing to the development of DGS technologies. The literature reviewed by Laborde et al. (2006) as well as other pieces provide numerous examples of interventions that use DGS in the context of construction tasks and that could serve as starting points for investigations on the conception of figure as construction. We take on the more specific focus on the means students have to control their constructions but look broadly across constructions with different instruments, sketching interventions that might strategically use those different instruments to develop particular means of control.

We explain now how we see the increased sophistication of the means of control in the context of our modeling approach. Students can do constructions, and the more they learn procedures and tools the more constructions they can do. The extent to which those constructions can represent figures is what is at stake: What are their means to control the correctness of statements in which they assert "this is a [name of figure]"? And, furthermore, what is the "this" that they can say is so? This question is particularly interesting because students' intellectual development is not linear. Early in life, children project their intention to their construction, but later they develop the capacity to judge their own construction on external standards and can recognize if their construction is not what they originally intended it to be (Piaget and Inhelder, 1956). Of course, that development is still too coarse to accommodate the distinctions between error (a misunderstanding of the properties of the figure) and inaccuracy (an unreliable execution of basically correct procedures).

The modeling approach applied to the construction conception means modeling geometrically the behavioral and physical activity of constructing small geometric objects. This means, as it did before, two things. On the one hand, it means creating a representation of construction: While, earlier, the construction was the artifact produced, now it will mean developing a more sophisticated sense of what construction means, particularly including the sense of construction as a reasoned procedure that is expected to produce an artifact with certain properties. On the other hand, it means using that representation to make anticipations about geometric figures, where the development will go from being bound to the experience of the constructed artifact (e.g., making an artifact and then looking empirically for what properties it has) to being able to (1) infer or calculate anticipations on the properties of the artifact based on the features with which the procedure will supposedly have endowed it and (2) intervene in a construction (e.g., creating auxiliary elements) and being able to interpret what is meaningful and what is idiosyncratic or ancillary in regard to how the artifact appears after such intervention. Overall, this modeling of the construction conception produces means to critique both resemblance as the criterion for success and bricolage as the method for success; it responds to those with a distinction between correctness and accuracy (or validity and reliability) as twin criteria for success, and legal reason and strategic reason (see Chazan and Sandow, 2011) as twin methods for success. In what follows we illustrate this with a few communication tasks that induce the modeling of construction as envisioned by these general comments.

Along the lines of conceptualizing the study of geometry as the mathematical modeling of students' experiences with shape and space, the various controls illustrated below can be seen as examples of how secondary geometry can serve the development of a control structure that helps students justify constructions as rational arguments for the existence of mathematical objects as opposed to constructions as manipulations of concrete materials for the creation of visual objects. Arguably the development of a control structure—that is, a system of dispositions-to-act as well as of accepted statements that underpin what is considered appropriate—is a step toward developing more deductive control structures.

It is clear, on the one hand, that constructing diagrams and artifacts in elementary school is different than in secondary school in terms of the role of tools and the preciseness of procedures. Early on, any tool and any use of that tool might be just as appropriate as any other (e.g., creating straight lines with a straightedge, with a steady hand, or with the line tool in DGS; making sure segments are constructed congruent by checking visually, or by measuring each) as long as the final product resembles what it is supposed to. At the secondary level, however, students have the opportunity to differentiate the procedure by which something is constructed from the eventually constructed object and interrogate the qualities of that procedure. Relevant in achieving such a distinction are reflections on (1) the mathematical underpinnings of the tool being used and (2) the reliability in the use of those tools. Constructions with straightedge and compass, such as the traditional construction of the perpendicular bisector of a given segment, provide basic raw material for developing such controls. If we imagine those traditional construction procedures being action tasks, a communication task posed apropos of them could bring about the reflection and critique desired.

5.4.4.1. Examining Reasons for Features of a Procedure

One aspect of constructions in which students can increase their means of control is by examining the reasons for features of a construction procedure. In particular, canonical construction procedures have some features that are necessary if the construction is to be correct, and other procedures that are just convenient for simplicity. Communication tasks could be studied in which students separate the former from the latter.

Consider, as an example, the traditional construction of the perpendicular bisector to a given segment. To remind the reader, the procedure is as follows: Given segment $\overline{AB}$, take the compass and open it so that it contains a segment whose length is more than half of AB; then, using that opening and centered at A, make a semicircle starting above A and ending below A; next, without changing the opening of the compass, make another semicircle centered in B and that starts above B and ends below B. Note the points at which the two semicircles meet, one above $\overline{AB}$ and another below it. Using a straightedge, draw the line that connects those two intersections; this line is the perpendicular bisector.

Now we invite the reader to consider the following scenario:

> A group of students (Julie, Mary, Peter, and Tom) carried out the construction of the perpendicular bisector as follows: Tom opened his compass centered on endpoint A of the segment $\overline{AB}$ and made an arc above $\overline{AB}$. Then Mary used that compass, also centered it on A, and made an arc below $\overline{AB}$. Then Peter opened his own compass, centered it on B and made an arc below $\overline{AB}$, and finally Julie used Peter's compass centered on B and made an arc above $\overline{AB}$. Julie's arc crossed Tom's arc at a point that Julie labeled X, and Peter's and Mary's arcs intersect at a point they labeled Y. Peter just drew line $\overleftrightarrow{XY}$ but it did not look like it intersected the segment at its midpoint.
>
> Another group of students (Dean, Juana, Nelly, and Seamus) did the construction as follows. Dean opened his compass centered on endpoint A of the segment $\overline{AB}$ and made an arc above $\overline{AB}$. Then Nelly used that compass, centered it on B, and made another arc above $\overline{AB}$. Then Seamus opened his own compass, centered it on A and made an arc below $\overline{AB}$, and finally Juana used Seamus's compass centered on B and made an arc below $\overline{AB}$. Seamus drew line $\overleftrightarrow{XY}$ but it did not look like it intersected the segment at its midpoint.

The two stories could possibly be presented to students using comic strips. Students could then be asked to indicate to each of those groups what they think may have gone wrong with each of these constructions, if anything. How could those groups of students improve their constructions?

We expect that such an analysis could address, in particular, the rationale for using the same opening of the compass for all the arcs, and how to ensure that the arcs will intersect. The notion that unless the opening of the compass is the same, the points of intersection of the arcs will not be on the angle bisector is connected with the theorem that every point on the perpendicular bisector is equidistant from the extremes of the segment, and is at the root of the stipulation that the two arcs above the segment need the same compass opening, as do the two arcs below the segment; yet one need not use the same compass opening above and below. As a consequence, the correct answer is that while the first group's construction is errorful (in that their procedure does not necessitate the perpendicular bisector), the second group's construction is only inaccurate (their procedure is correct though they may have made some sloppy mistake). The necessity of the arcs to intersect could also be raised with a comparable scenario. This sort of critique of failed constructions can help bring out the theoretical underpinnings of a construction. Further construction tasks can suggest communication tasks in which critique and reflection on construction procedures can achieve mathematically valuable ends.

5.4.4.2. Gaining Theoretical Control of Initial Conditions

A second issue on which communication tasks could help develop students' means to control constructions has to do with the degrees of freedom available in the choice of givens for a construction, and is a key element of difference between the construction and description conceptions of figure, at least at the secondary level. Note that when geometric figures are described in terms of their properties many properties may appear concomitant. The work of proving that some of those properties entail others (for example as shown in the section above with the use of the game "Guess my Quadrilateral") contributes to the deconstruction of such concomitance of properties, inducing an implicative structure. Yet that implicative structure among properties only takes care of part of the study of geometric objects: It allows one to say that if an object exists that has a property, then, by virtue of those implications, such an object will also have other implied properties.

What is missing there, of course, is the extent to which one can really assert that an object with the initial set of such properties exists. The story of Saccheri's quadrilateral (which can be defined to be more general than rectangles in Euclidean geometry, but which cannot be proven to exist as a class different than that of rectangles in Euclidean geometry; Bonola, 1955) can be handy here for realizing that a figure does not exist merely by postulation. We contend that construction activities, particularly communication tasks associated to construction tasks, can provide students with perspective and opportunity to consider critically and reflectively the possibility of existence of geometric objects with postulated (i.e., given, initial) conditions.

Consider the construction of the circle tangent to given lines at given points of tangency, as suggested in the animation The Tangent Circle[5] (see description in Herbst & Miyakawa, 2008; also Herbst, 2008; Herbst & Dimmel, 2011). In this animation, a geometry class is asked to create a circle tangent to two given intersecting lines, at two given intersecting points. Students are presented with the diagram below on a sheet of paper; also, orally the teacher tells them "In the figure there are two lines *a* and *b* and they intersect at point *P*. You'll also see that I've marked a point *A* on line *a* and a point *B* on line *b*. I want you to draw a circle that is tangent to *a* at *A* and tangent to *b* at *B*" (see Figure 5.10)

As noted by Herbst and Dimmel (2011), this task is ambiguous with respect to its initial conditions. On the one hand, as represented verbally, the task characterizes the points *A* and *B* merely as being on the corresponding lines; in Laborde's (2005) terms, the theoretical property of each of those points is that they belong in each of those lines. On the other hand, the given diagram represents *A* and *B* at particular places on the lines: Spatiographically, the points are at given distances from *P*, and those distances happen to be different from each other. In this case the spatiographical property that the distances *AP* and *BP* are unequal is related to a theoretical property of the successful construction to be done: Namely, that the construction of a circle tangent at both points is possible only when those distances are the same.

Note how the ambiguity of those conflicting properties is heightened by the fact that segments $\overline{PA}$ and $\overline{PB}$ are not brought to existence explicitly; rather *P*, *A*, and *B* allow for $\overline{PA}$ and $\overline{PB}$ to be seen as segments (in Coliva's, 2012, sense), but also to be seen as lines. And that muted choice of how to see the points

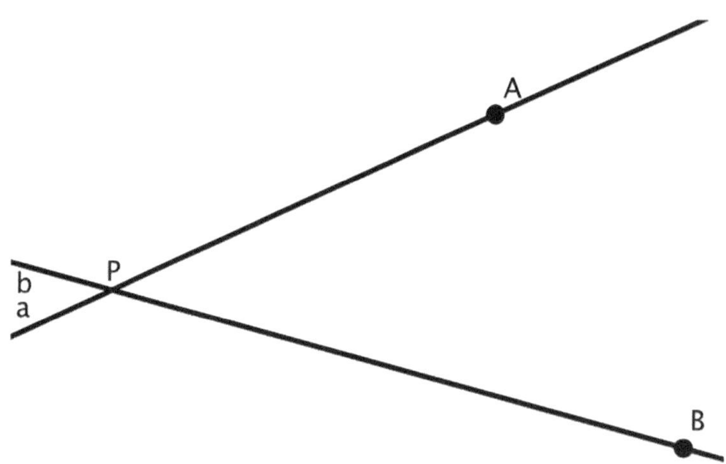

FIGURE 5.10 Draw a circle tangent to *a* at *A* and to *b* at *B*

correlates with the extent to which traditional Euclidean ways of reading the diagram (in Manders', 2008, terms, whether the diagram shows exact or co-exact properties; see Chapter 2) are activated: If one sees only lines, then *A* and *B* are only incident in lines, and incidence is an appropriate, co-exact, property that one customarily reads off the diagram; but if one sees segments as well, then *PA* and *PB* are lengths, and their inequality is not an appropriate property to read off the diagram (since congruence is, in Manders', 2008, terms an exact property).

In other words, the Euclidean standard does not allow us to assume that *B* and *A* are settled at their apparent locations; but of course, students do not know that, and the statement of the task (the circle needs to end up being tangent at *A* and at *B*) treats both the theoretical and spatiographical points as the same. The action task, to construct the circle, entails a breach of the didactical contract one way or another: Either students are being asked to do something they cannot possibly succeed at, or they are being expected to ignore information that is presented to them as relevant and change it, which is something they do not often do. The payoff of this breach of contract is, however, justifiable; the animation The Tangent Circle illustrates how this action task might lead to an interesting discussion of the tangent segments theorem.

Clearly, secondary school students could be asked to do this construction in much the same way that students in the animation were, or they could be shown the animation up until a moment when a student (Lambda) says that nobody could do the construction unless they could move the points, and the animated teacher poses a second question. Or students could be engaged in a communication task, like the second question the teacher in the animation posed to Lambda: Once two lines are drawn and one point of intersection is plotted, the teacher asks Lambda to indicate where the second point should be plotted in order for the tangent circle to be constructible.

Students could then be asked to locate the points of tangency so that the construction could be achieved, and asked to specify how to do the construction. A task like this can help students understand that the possibility of a construction depends on the initial conditions; that is, that initial conditions could be such that the construction is impossible. Likewise initial conditions could be such that the construction still requires specifying more givens, as would be the case if the question had merely been to construct a circle tangent to the two lines (and no other requirement was established).

5.4.4.3. Turning Declarative Statements into Construction Operators

The Tangent Circle construction also brings to mind an important element of the work of constructing and its relationship with prior knowledge. Apropos of his own use of a variation of the tangent circle construction, Schoenfeld (1989) had noted that students who had proved the theorems about tangents to circles could still behave as empiricists when dealing with construction problems, making no use of the knowledge contained in those theorems. In his studies of students' production of proofs, Anderson (1982) had noted, however, that elements of declarative knowledge (i.e. propositions, theorems) need to be transformed (compiled) before students can use them as action rules. Anderson (1982) made this point in relation to the use of congruence criteria in triangle proofs; students had to transform statements of the conditions upon which two triangles could be said to be congruent (e.g., if two triangles have their corresponding sides congruent, they are congruent, that is, their corresponding angles are also congruent) into action rules that indicated what they had to do to establish the congruence of the triangles (e.g., find pairs of corresponding sides and determine whether they are congruent, then claim that the triangles are congruent). In other words, even in problems to prove or to find, knowledge needs to be transformed in order to be operational. The conceptions model proposed by Balacheff and Gaudin (2010) contains that distinction in the notions of *control* and *operator*; while a proposition says what is or should be true, and can serve to control a state of affairs, an operator indicates what one should do, and can be used to produce a state of affairs. In construction problems this need for knowledge to be transformed from declarative statement to action rule is just as relevant.

In the case of the construction of the tangent circle one could see the same transition being needed between the declarative statement that a line tangent to a circle is perpendicular to the radius at the point of tangency and the action rule that to find the center of the circle one can restrict oneself to finding a point on the line perpendicular to the given line at the point of tangency. The same can be observed as one considers the 'division of the land' problem of a Japanese lesson that Stigler and Hiebert (1999) analyzed (see Figure 5.11; see also Koschmann and Mori, 2016): In this lesson, students had first been reminded that they had studied how triangles of the same base and whose third vertex lies on a parallel to the base have all the same area (Figure 5.11a), the students were then asked to consider two neighbors (Eda and Azusa) whose properties were separated by a crooked line and asked to propose where to place a straight division (Figure 5.11b). The lesson shows that students did not instantly know what to do, even though once the construction was done, it could be seen merely as an application of their prior knowledge (as the teacher in the lesson says, "just turn your head around"; Koschmann & Mori, 2016; for more about Japanese lessons, see Fujita and Jones, 2014). We hold the Japanese lesson as a valuable example of work on an action task that could help develop the transition from declarative knowledge to action rule. The extent to which such transformation is explicit, the extent to which the activity actually produces an operator rather than just an accomplished operation, could be worked out with a communication task.

Consider the following variation of the lesson. Present the division of the land problem without any reminder of what had been studied before, divide the class into two teams, and call one student from each team to the board to solve the problem in any way they can. In circumstances like those, we have observed students[6] doing things like (1) closing the areas, calculating the areas with the old

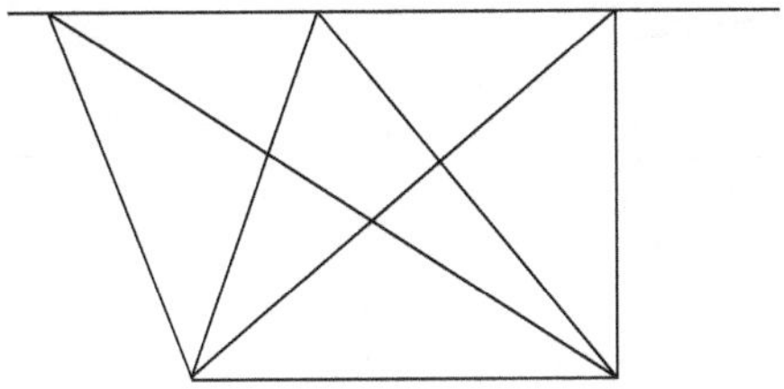

FIGURE 5.11A Reminder of theorem studied in prior class

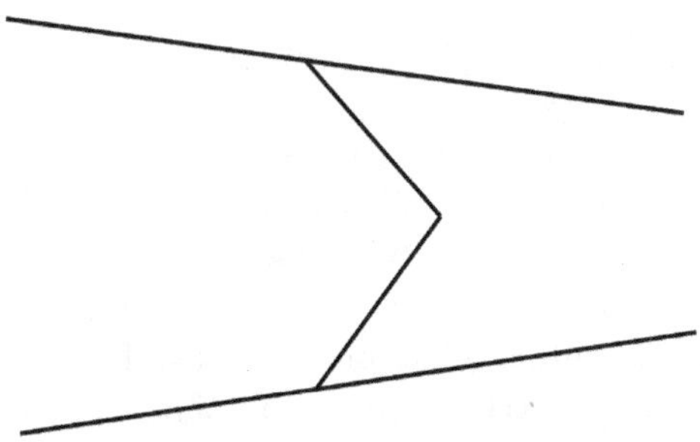

FIGURE 5.11B Division of the land problem

boundary, proposing a new straight boundary, calculating the areas for that and adjusting continuously (Figure 5.12a; vertical lines are the lines students draw); and (2) drawing a line that passes through the midpoints of the crooked sides, checking the equivalence of the areas traded, then adjusting continuously (Figure 5.12b). We present this here to suggest that students have some empirical ways of thinking of the problem that might make it possible to pose this as an instance of a situation of construction. The real communication task comes after.

The teacher could give students in both groups (communicators) a paper statement of the property ("if two triangles have the same base and the vertices

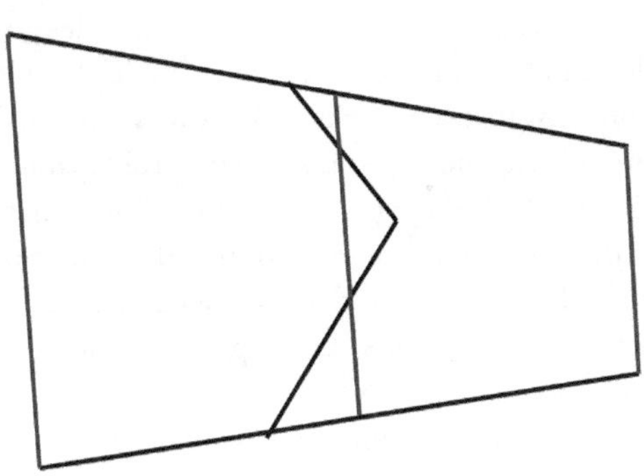

FIGURE 5.12A Observed solution equalizing the actual areas

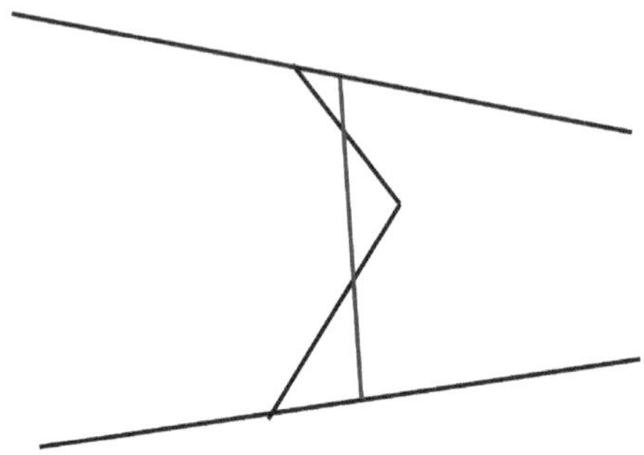

FIGURE 5.12B Observed solution equalizing the traded areas

opposing the base are contained in the same parallel to the base, then their areas are equal") and ask them to direct the student at the board on how to solve the original problem without the need to calculate and approximate. The communicators would have to give explicit instructions of what the student at the board needs to do but not read off the property. One could expect such work to produce instructions like: Draw a line connecting those two points, draw a parallel that passes through the third point, etc. Obviously there is no assurance that such work will produce the correct procedure from mere adaptation to the milieu. We leave this at this point and suggest to the reader two possible paths: (1) to design the communication task in such a way that students interact with an adidactical milieu or (2) experiment with the task as posed and investigate what the teacher does to produce the communications desired.

5.4.4.4. A Declarative Statement and Its Associated Action Rules

Construction tasks offer the opportunity to learn how theorems participate in an economy of knowledge preservation. Indeed, the distinction made above between theorems as declarative statements being transformed, in the context of construction tasks, into action rules or operators, highlights the notion that theorems contain more than one meaning. In fact, as far as action rules go, we might also say that theorems could contain multiple action rules, multiple operational meanings. For example, the theorem that if a line is tangent to a circle it is perpendicular to the radius at the point of tangency, entails the action rule that to find the circle tangent to a line at a given point of tangency one should draw the line perpendicular to the given line at the point of tangency and pick the center of the circle on that line. But the theorem also entails that if one has a circle and a point on the circle, to draw the tangent line to the circle at that point one just needs to draw a line perpendicular to the radius at that point.

Awareness of the multiplicity of meanings associated with a theorem could be raised with communication tasks such as those in which students are to reproduce a figure that they cannot see. Suppose for example that two teams (each composed of

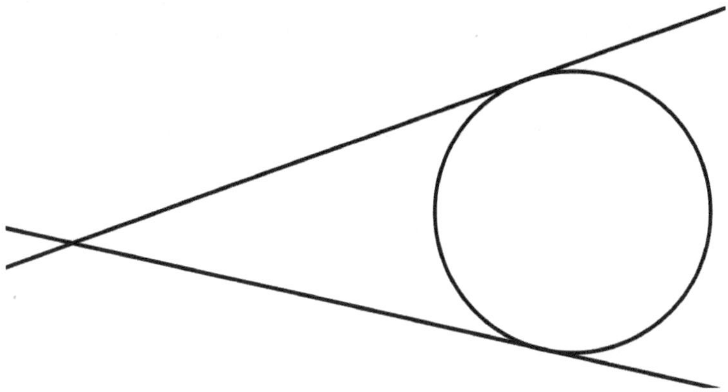

FIGURE 5.13 Giving directions to reproduce this diagram

producers and receivers of messages) need to reproduce the same diagram or model, for example, the diagram seen in Figure 5.13. The producers have access to the complete diagram while the receivers cannot see it. The receivers from one team have only a copy of the circle, while the receivers from the other team have only a copy of (one of) the lines. The producers might also have available the statement of the theorem considered above (the line tangent to a circle is perpendicular to the radius at the point of tangency) and they have to give verbal instructions to the receiver to create a diagram that can be superimposed with (or projected on top of) the other. A discussion could ensue on how both groups used the theorem, making explicit the different action rules and how they are entailed by the same theorem.

5.4.4.5. Summing Up This Section

The four preceding subsections have been meant to illustrate how communication tasks could be used to help students extend the set of controls they use for their work constructing representations of figure, hence helping to develop the conception of figure as construction of objects in the microspace. The subsections have not been organized in response to geometry topics, but rather in terms of the various ways in which the controls of a conception might become more theoretical. In the first example we illustrated how students could be engaged in questioning the necessity of various features of a construction procedure, realizing that while some of those features must be used, others are only convenient to use. In the second we noted how communication tasks might be used to gain theoretical control of initial conditions for a construction. In the third subsection we illustrated how students might be scaffolded in turning declarative statements into action rules, and in the last subsection we illustrated how students might be scaffolded to derive multiple action rules from the same declarative statement. These four are only examples of how the work in secondary geometry classrooms could promote more theoretical control and operations on the conception of figure as construction of objects in the microspace. Without a doubt these examples can be extended to cover other issues that are pertinent to the work of modeling mathematically the students' experience with shape and space (in this case the experience of making shapes). These and other issues connected with helping students increase theoretical control of their construction activities could be promoted with communication tasks; our community needs more classroom research oriented to turning elements of the transition to theoretical geometry into instructional tasks where students can learn to act like (pure and applied!) mathematicians.

5.5. Conclusion

The intent of this chapter has been to provide concrete examples of how a modeling approach to geometry instruction can be instantiated in specific activities that extend students' conceptions of figure. The notion of modeling as constructing operational representations (Jahnke, 2007) was applied to the four conceptions of figure that we had introduced in Chapter 2 (figure as navigation of the macrospace, figure as capture of an object in the mesospace, figure as description and manipulation of an object in the microspace, figure as construction of an object in the microspace). In each example activity, one of those conceptions of figure was the target of the modeling while another was used to build a model that avowedly could serve to bring students to a higher level within a learning trajectory for the conception of figure being modeled. In each example, a notion of informal proof supported the operational capacity of models—the sense in which they can assist the reasoned prediction of what will be true in the modeled world. In each example, the notion of proof was informal in the sense of Lakatos (1976), that is, it did not require well defined concepts and axiomatized domains to operate, but rather it operated with representations of concepts (e.g., "box") that could progressively be endowed with characteristics and related to other representations by way of increasingly explicit assumptions. Clearly, these activities do not yet constitute a curriculum but merely illustrate a perspective from which a curriculum might be developed, as well as create contexts in which to develop research on geometry learning and geometry instruction. The activities proposed in this chapter may be used to research the learning trajectories that secondary students follow as they progress in their understanding of geometric figures. The four conceptions of figure presented in Chapter 2 not only help organize a large space in which students' learning of geometry is to develop, but also help assimilate a point documented by developmental researchers who studied young children's learning of geometry, reviewed in Chapter 3: That students might be at one level in some ideas and at another level in other ideas, thus creating the need to speak about learning trajectories that are specific to mathematical goals. In our case, microspace conceptions are likely to be further along in their development than macro or mesospace conceptions. The activities proposed help pose a question for research on geometry learning: To what extent do the characteristics of existing conceptions of figure used to model students' experiences with shape and space support or constrain their development of more sophisticated conceptions of those experiences? Specifically, as the microspace conceptions are brought to model students' experiences in the macro and meso spaces, how do they facilitate or constrain students' capacity to predict and explain those experiences? The same questions can be asked about students' use of existing microspace conceptions to develop more advanced microspace conceptions. Finally, the activities proposed in this chapter can also be used to research the teaching of geometry as a subject-specific and situation-specific work. In order to organize those activities, the teacher of geometry needs to make use of existing sociotechnical structures, such as the didactical contracts and instructional situations described in Chapter 4. Those bring with them expectations for what the teacher is supposed to do and what students are supposed to do that can also support or constrain the work they will do together in these activities. Readers who are interested in doing research on the teaching of geometry can therefore ask how those sociotechnical structures mediate the work students and teacher do in these activities. And if these activities demand that teacher and students negotiate and possibly discard some of those sociotechnical structures, on what grounds are such negotiations justified from the perspective of the practitioner. These questions for research on the learning and the teaching of geometry become central to our discussion as we move to the last chapter of the book.

Notes

1 The informed reader should realize that what we refer as the student's "mathematical task" includes what Brousseau calls "a-didactical situation". We reserve the word "situation" to its use in the context of instructional situations; hence we refer to the various kinds of a-didactical situations from the TDS (e.g., action, communication) as "tasks"—e.g., action task, communication task.

2 We chose Resistencia because its layout has been designed with enough attention to geometry to make this task simple while interesting. Other cities could be used alternatively.

3 By saying that a representation is analytic we mean that it realizes a whole as an organization of elements that can be selectively taken apart and combined with others so as to make arguments: The labeling of the vertices of a square A, B, C, and D, for example permits talking about the square $ABCD$ as well as of side $\overline{AB}$, of vertices A and C, and even of a not-yet-drawn diagonal $\overline{AC}$. Labeling the vertices turns the drawing of a square into an analytic representation. Duval (2005) calls the cognitive capability that realizes such analytic representation "dimensional deconstruction".

4 L'articulation cognitive entre le registre de la visualisation et celui du langage ne se fait pas au niveau des mots mais à celui des propositions.

5 The animation can be seen among the elements of the Original Collection in LessonSketch, www.lessonsketch.org

6 Our observations have been informal, both with middle and high school students and with preservice secondary mathematics teachers.

6

A CONCLUSION AND A BEGINNING

Doing Research on the Teaching and Learning of Secondary Geometry

6.1. Introduction

We wrote this book for graduate students in a Masters or doctoral program, either practicing teachers seeking to know more advanced material to do research in their practice or incoming academics seeking to start a career in mathematics education scholarship. Up to this point the book has reviewed theory and research and has suggested ideas for interventions that could be developed further and implemented in secondary classrooms. The thought was that Masters' students might want to use those interventions as context for research that could lead to a Masters' thesis; likewise beginning doctoral students might use these ideas to undertake intermediate projects that might eventually lead to a dissertation. Along those lines, this concluding chapter begins with a discussion of some research questions that could be asked about these interventions. We address this discussion to graduate students in general, knowing that some questions will appear more interesting to practicing teachers who do research in their practice; others will appeal to Masters' students in search of a thesis project; and yet others will seem productive for longer-term projects.

6.2. Research Questions

What might a graduate student do with this set of communication tasks that we have sketched? Hopefully, we have sketched them with enough detail to provoke the readers to think about what these tasks might look like when implemented in a real classroom. Hopefully also, our discussion in Chapter 4 about instructional exchanges and instructional situations has contributed to impress upon the reader that these communication tasks are likely to be novel, unfamiliar in the instructional situations normally found in secondary geometry classrooms. That observation, of course, neither means the tasks are unworthy of consideration nor that instructional situations are useless tools to think about geometry instruction. Rather, what we have presented thus far is meant to be the basis upon which to mount research and development projects that could be of interest to the reader.

In what follows we integrate the material presented in the previous chapters to suggest an agenda for research that looks at these communication tasks both as loci for research on students' thinking and learning, and for research on teaching and teacher knowledge. Rather than develop the themes of this agenda for each communication task, we do that in general here and suggest that the reader undertake working through the details for specific tasks.

6.2.1. Communication Tasks as Design Research Projects

Given any of these communication tasks, one genre of questions to ask is related to fleshing them out into curricular interventions that create opportunities to learn and to examine the extent to which such opportunities to learn are fulfilled. The *design experiment* has been aptly described as a research design in a number of pieces including Cobb (2000) and Cobb et al. (2003; see also McKenney & Reeves, 2013). The French tradition of didactical engineering (Artigue, 1994), the Japanese tradition of Lesson Study (Lewis, 2002), as well as design approaches used in Germany (Wittmann, 1995) and the Netherlands (Van den Akker, et al., 2006) contain examples of how researchers have engaged in an iterative process of constructing classroom interventions and studying the learning opportunities they create, including possibly the learning trajectories that become visible in those contexts.

For each of these communication tasks (or perhaps for a set of them) one needs to ask two broad, related questions. One question is what a curricular intervention looks like that includes these communication tasks. The second is what student learning looks like in the context of those tasks. In regard to the first question—what the curricular intervention looks like—this can be unpacked further into two sets of questions: (1) a local design question, that requires one to think about the specific lesson or lessons where the task will be deployed, and (2) a global design question, that requires one to think about larger curricular units in which those lessons will fit.

The local design question will lead the designer to analyze a priori what each of the tasks in the lesson could involve. In Brousseau's (1997) terms this would include identifying the milieu, the student's actions on the milieu, and how the milieu may respond to the student's actions. To be clear, this requires the designer to do two sorts of anticipations: One that characterizes the task from the point of view of what students will explicitly be asked to do, and another that spells out what the designer can anticipate the student could end up doing. These are not equivalent. The local design that seeks to produce such anticipations can benefit from considering the various components of a task. Doyle (1988) proposes that an academic task for students can be described in terms of the avowed goal or product, the resources available, the operations anticipated, and the value assigned to the work in the accountability system of the class.

To illustrate how this local design analysis could be done we recall the task used by Herbst (2003), in which students were asked to collaborate in ranking a set of eight triangles according to area. The goal given to students was for them to come up with a list of the triangles sorted according to area and the reasons for each comparison. The operations were not explicitly indicated, but once the teacher had ascertained that students remembered how to calculate the area of a triangle, he asked them to use the area formula as little as possible. The resources included the triangles, that had been cut out in cardstock and were rather small (all triangles were smaller than one quarter the size of the palm of a hand), and a set of measuring tools and calculator. There was no apparent grade associated with completing the task, but the task was presented as a problem of the day in a problem-based class, so participation was crucial. But from the teacher's perspective, as shown by Herbst (2003), the task was slightly different. First, the goal was for the students to come up with basic properties of the area function (e.g., inclusion,[1] additivity[2]) and use them to compare triangles (writing versions of these basic properties as the reasons for their comparisons). In particular, a goal of the task was to discover the reason why two triangles (*D* and *E*), which had been constructed so that neither of them was completely included in the other, were in fact equal in area. Second, while the operation of actually calculating an area was discouraged, students were expected to come sufficiently close to using it so that they could realize that the area formula could *express* an area rather than merely produce it (e.g., $\frac{1}{2} \times b \times h$ not only indicates a calculation to be done to produce the area when the numbers are multiplied through, but also, and more importantly, it represents the area as an algebraic expression; that is, $\frac{1}{2} \times b \times h$ *is* the area, not just a prescription for how to find the area). Third, the triangles on cardstock cutouts were meant to represent idealized triangles and hence be taken as triangles in spite of the fact that they were in fact three-dimensional objects (triangular prisms, since the cardstock has thickness) and the way the cuts had been realized might make their bases more like polygons than triangles. Finally, for the teacher what was at stake was the students' development of basic ideas of the area function, including among others that if the base of one triangle is half that of another but the height of the first is twice the height of the second, the two triangles are equal in area. As one can tell from this brief comparison, the overt task for the student may look quite different to the covert task that the teacher had in store for the student. The designer of any communication task needs to have in mind both what the task will apparently be like for students and what the task is going to enable the teacher to observe.

Furthermore, while one task may be the centerpiece of the lesson, the work of getting the students prepared to do that task may involve other tasks. In the example from Herbst (2003) the teacher began by asking the class whether they remembered how they had learned to calculate the area of a triangle, and in that context he made sure that students knew what "base" and "height" meant. Thus, the involvement of students in the designed task might be eased through earlier work framed using the resources of familiar tasks, particularly of tasks that belonged to existing instructional situations. On the other hand while the knowledge activated through that earlier work might facilitate the engagement in the target task, it might also activate situational norms that the new task might need to breach. Such is the case with the question about the area formula in the example: While the initial work allowed the teacher to make sure students knew what the area formula for triangles was, it also brought up for students a normative meaning of what it meant for them to "use" the area formula.

Finally, once a task has been launched and possibly gotten to a point where it has enabled the work envisioned, the teacher needs other tasks to complete the lesson. In the example provided, once students had produced a completely ordered list of triangles, the teacher might have asked the students to write as general properties the reasons they had used to decide whether a triangle had less area than another or equal area to another one. In sum, the work of designing tasks requires working inside each task, identifying its components from the perspective of the student and the teacher; the work also requires designing neighboring tasks that prepare students to do the focus task or that help students process what has been done along with criteria that the teacher may use to transition from one to the next. And, as noted before, it requires working at the broader scale of scope and sequence, worrying about what came before and what comes after in the coursework.

The work of design research involves using experience as part of the process with which these sequences of tasks are designed. The designer may have some initial ideas about what students might be given as resources and asked to do, and some expectations as to the work that might ensue. Observation of students at work in such conditions may reveal needs that had not been anticipated in advance (e.g., extra resources, earlier or later tasks, more supports for the teacher), which could be incorporated in newer versions of the intervention. The iterative nature of design research could, over time, help develop learning trajectories associated to the concepts or processes being handled, particularly with the processes of mathematical modeling of students' experiences with shape and space.

The work of iteratively designing an intervention, using experience with a version of the intervention as a basis for tinkering with its design, highlights the second, broad question listed above: The designer needs to anticipate and observe what students' thinking and learning might look like in the context of the intervention. As noted earlier in this book, this work may include identifying the conception of figure that might be summoned by the tasks in which students are to engage, identifying observables in the activity of the student that might attest to transitions to a different conception, predicting errors that students might make when trying to assimilate a new task to an existing conception, and so on. Herbst (2005) shows how an intervention designed around another area of triangles problem[3] initially appealed to a very restrictive conception of area equivalence (namely, that congruent triangles have the same area) but later created an opportunity to observe students' adaptation of that conception into a more powerful one (that if two triangles can be divided into figures that are respectively congruent, the two triangles have the same area). Clearly, one option for researchers using these designed environments is to explore how students engage their prior conceptions and how the tasks perturb those prior conceptions: What new conceptions emerge from students' engagement in the new tasks? Hopefully the material of Chapter 3 has now become meaningful in a different way: While in the reader's earlier passage through Chapter 3 that material might have been seen as information about students' thinking and learning, the present chapter should provide context and impetus for the reader to review Chapter 3 as examples of what one could be asking questions about and how one could frame the observations one can make about students' thinking and learning.

While from a curricular standpoint, the work of design research involves using experience to refine the characteristics of an intervention, from a study of mathematical thinking and learning the work of design research involves describing the various conceptions that students engage as they work in the tasks that are part of the intervention. Quite a bit of mathematics education research on students' thinking and learning is routinely done in the context of designed interventions. Both of these perspectives—the use of students' experience to refine the design and study how the tasks work, and the use of a designed intervention to explore students' mathematical thinking and learning—are complementary in design research and they describe what readers might do with the tasks that have been sketched earlier in this chapter.

6.2.2. Communication Tasks as Instructional Experiments

A second genre of questions to ask apropos of the use of these communication tasks, as well as other interventions, has to do with the work of teaching involved in them. With this we refer not so much to the way in which the design process prescribes some of what the teacher will do (e.g., choice of tasks and how they are worded, etc.) but rather to the work a teacher has to do or ends up doing to manage the work that students happen to do in those tasks. Much has been written in the last 30 years about the study of the work of teaching in instruction, and it will not be possible to summarize it here; some work has also been done specifically on the teaching of secondary geometry (reviewed in Chapter 4) and more can be learned from handbook chapters that discuss recent research on mathematics teaching (Chazan et al., 2016; Franke et al., 2007; Herbst & Chazan, in press; Hiebert & Grouws, 2007). Here we make just a few comments to suggest the kinds of research in which readers could engage as it relates to these communication tasks. Because the readers of this book include practicing teachers, it may be worth noting that a teacher could engage in this research by looking at her or his own practice too. Ball (2000) has conceptualized this kind of design research as "first person research"; Chazan (2000) and Lampert (2001) have shown how this particular kind of research on one's own teaching can be written to disseminate knowledge about teaching. We could learn quite a bit about the work of teaching from reading the first person accounts of a teacher who endeavors to engage her students in tasks like the ones we have sketched in Chapter 5.

A general perspective with which to look at this research is the notion that the use of novel tasks is a kind of instructional experimentation. It is experimentation in the sense that it systematically perturbs the natural conditions in which instruction takes place and creates, as a result, conditions to observe how instruction responds to such perturbations; that sense of experiment (also known as demonstration) has a noble genealogy in scholarly research, starting with Francis Bacon. They are not, of course, experiments in the modern, social science sense of random assignment of participants to conditions. Herbst (2006) described these instructional experiments by analogy with the breaching experiments used in the late 60s by sociologists interested in understanding how social groups would manage alterations to the social order. We consider each classroom to be an instance of instruction[4] (a system of interactions among teacher, students, and the knowledge at stake) and that such a system has customary ways of organizing those interactions. As we note in chapter 4, each class has a didactical contract that regulates in general what it means to teach and to study mathematics; it also has instructional situations that specify the operation of such a contract for particular kinds of knowledge at stake (Herbst and Chazan, 2012). Some sense of what is normal or unremarkable exists in any class, including what happens through the year, through a day, and in particular types of familiar tasks. Novel tasks, such as those sketched in Chapter 5, can challenge that sense of normalcy, not only for students who may get the opportunity to develop new knowledge or challenge old conceptions because of the novel task, but also for teachers who might see their work environments altered.

Clearly some of those alterations of the teacher's work environment are scoped in advance by the design of the task, but others are necessarily emergent, and it is important to document what those are. To do so may help understand in general the sort of challenges tasks might go through when they are implemented with students (Stein et al., 1996; see also Henningsen and Stein, 1997); it may also help understand the demands tasks put on teachers' knowledge. Herbst and Chazan (2012; see also Chazan et al., 2016) articulate the practical rationality of teaching as emergent from the interplay between three sets of elements: (1) the norms of the activity systems that a teacher customarily participates in, such as didactical contracts and instructional situations; (2) the individual assets of a teacher (such as a his or her mathematical knowledge for teaching or his or her beliefs); and (3) a set of professional obligations to the discipline, to individual students, to the class as a social group, and to the institution of schooling (see Chapter 4). We suggest that these elements can be used to inspect how a novel task, or a lesson including a novel task, might perturb the work of the teacher.

6.2.2.1. Researching the Behavior of Activity Systems such as Contract, Situation, and Task

A novel task, to the extent that it may draw upon norms of customary activity systems to be deployed, might also create tensions associated with the norms of those activity systems that are breached. The norms that might be breached could be either norms of the contract or norms of an instructional situation. Consider an example from the didactical contract first. Every novel task that is assigned to the class will rely on some norms of the didactical contract in order to be deployed in the class. For example, one clause of many didactical contracts is that the teacher has the right to assign work for students to do; this one is usually complemented with the norm that students can expect to be able to do what the teacher asks them to, and to do so by using what they have learned before. A problem like the one presented initially in the animation The Tangent Circle (where students are asked to construct a circle tangent to two intersecting lines at two points that appear not to be equidistant from the intersection; see Chapter 5) can be seen as a breach of this norm: The teacher is asking students to do something that turns out to be impossible for them to do without altering the conditions of the problem. While doing this may be justifiable (see Herbst & Chazan, 2012; also Herbst & Dimmel, 2011; Herbst, 2008), it can also be risky: Teachers who have seen that animation have often reported that they wondered whether the next time a teacher poses a problem that appears difficult students will refuse to work earnestly on it on account that it, too, might be an impossible one.

Additionally, a novel task might rely upon existing instructional situations in order to be deployed, and in so doing it might also elicit tensions for a teacher. Consider the task of asking students to prove that the angle bisectors of a kite meet at a point. At least in the United States, such a task's use of the word "prove" would summon expectations related to the instructional situation "doing proofs" (see Herbst et al., 2009), including, for example, that students will write in a two-column form a sequence of statements about a diagram of a kite and reasons that justify those statements on account of theorems, postulates, definitions, and the givens. To make that work possible, usually the teacher provides a diagram in which, among other things, all auxiliary lines are provided and all points to be used are labeled; the statement of the problem would also, customarily, be made in terms of a particular diagram (i.e., referring to geometric objects by their labels rather than to geometric concepts by their names; see Herbst et al., 2013). A teacher could breach these norms by stating the problem in terms of general concepts and expecting students to draw the diagram. But then some tensions could surface: If students were not to draw the diagram with some degree of accuracy they might be unable to use it to identify what triangles might need to be proven congruent (Herbst, 2004), but if they did draw the diagram with accuracy they might fail to see that anything needs to be proven. A teacher might be able to anticipate the difficulty and possibly transform the problem to a normal version, but he or she might also go along with the task as suggested and have to deal with students' puzzlement. Only some of such puzzlement could be handled by design, for example by providing guidance to the teacher on how to handle the different ways in which students might draw the diagram. The teacher would still have to handle much of that puzzlement on the spot as he or she manages interpersonal relationships with and among the class.

Along those lines, a question for research on teaching is thus to describe the ways in which teacher and students handle breaches of norms of the contract or of the situation as they negotiate how to work on a novel communication task such as those sketched in Chapter 5. Researchers interested in a particular novel task, such as one of those communication tasks sketched in Chapter 5, could use the designs to anticipate aspects of the task or moments in the enactment of the task where breaches of norms of contract or situation might be observed. They could then observe records of the enactment of such a lesson for evidence of whether and how teacher and students interacted at those moments. The need to maintain the engagement of students in the didactical contract may be used to explain the possible observation that the teacher might still alter the designed lesson in some way, or that the students may demand such alterations. The sort of work that emerges from such negotiation is thus likely to be not merely an enactment of the design but an adaptation that may or may not maintain the task's cognitive demands (Stein, et al., 1996). That is, a third element that might be described is the actual mathematics emerging from those negotiations. As a result of whatever modifications may be made to the students' scope of work to make it viable, they may end up doing mathematical work that differs in some way from what designers envisioned. Describing what such mathematics consists of is a key piece of this kind of research.

6.2.2.2. Researching Teachers' Individual Assets Such as Beliefs and Knowledge

A novel task, to the extent that it calls a teacher to enact instruction they may not be used to, to observe a multiplicity of mathematical responses on the part of students, and to figure out ways to respond to unexpected events, might also challenge the individual assets of the teacher. Within some institutional boundaries, most teachers are able to enact a personal style of teaching as well as to profess a range of beliefs about mathematics instruction (Chazan, et al., 2016). One could expect that enacting a novel task might be, for some individuals, somewhat of a stretch from their usual teaching style, or even require them to attach value to mathematical behaviors that they don't usually value. Furthermore, managing students' work on a novel task may challenge the teacher's specialized content knowledge (for example, as he or she tries to make mathematical sense of non-standard productions by students), the teacher's knowledge of content and students (for example, as he or she tries to ascertain what students may be missing when their answers are incorrect), or the teacher's knowledge of content and teaching (for example as he or she decides how to scaffold the task for someone who is not able to start working with what the original task statement provided; see Herbst and Kosko, 2014; Kuzniak and Rauscher, 2011).

Novice researchers could take on the work of researching the demands that novel tasks make on a teacher's personal resources as well as the personal strengths that such tasks might call forth. Researchers could strive to account for those, for example by using interviews about lesson plans before the lesson, observation of teacher behavior during the lesson, and video-stimulated recall of the lesson taught in the context of post-lesson interviews with individual teachers. This research can complement observations of the design of lessons, by informing on the extent to which lessons accommodate the individual differences among teachers. While research on teacher beliefs has often relied on single case studies, we recommend that when this kind of focus is applied to designed tasks researchers try to incorporate multiple case studies of the "same" lessons so as to be able to see individual differences against the background of task commonalities.

6.2.2.3. Researching the Sources of Justification for Actions in Teaching

Since the enactment of novel tasks is likely to require negotiations between teacher and students, even in the event that teaching actions followed the design to the letter these actions could not be lightly accounted to fidelity to design (Confrey & Stohl, 2004, p. 114). Rather, such actions, compliant or not, need to be considered as deliberate, resulting from actual or implied assessments of viability, particularly by the teacher. The teacher and the students may agree to do things that they personally like as well as things that they personally might not like; in either case it is important to ascertain the grounds on which they might find those actions justifiable.

Herbst and Chazan (2012) have proposed four sets of obligations as possible sources of justification for teaching actions. These include an obligation to the discipline of mathematics, an obligation to students as individuals, an obligation to the class as a social group, and an obligation to the institutions of schooling. These obligations are likely to be muted in intact teaching, in the sense that when nothing special is being done, the actions taken may find their way through by force of custom; that is, while they might be justifiable, those justifications may be hard to retrieve explicitly. Hence, while actions in intact teaching may respond to those obligations in some way, their justifications might not be easy to identify; nevertheless those actions are likely to feel to the practitioner like they make sense. But in specially designed interventions, like the ones scoped in Chapter 5, regardless of whether they are or are not successful, we might find prime opportunities to take those obligations out of such dormant state. A key goal for this research is to elicit from teachers the particular grounds on which various decisions and actions in enacting an intervention might be justifiable. Those actions may include restating a task to include something that had not been planned, deciding to bring one student instead of another to the board, deciding to transition from one task to the next, etc. They are all examples of actions that might have been viable in the enactment of an intervention. The question is why they were justifiable or why it would make sense for a professional to take them on; those questions can be asked and responses can be used to better articulate the grounds of justification for actions in teaching. While the actions themselves and their own specific justifications may be too specific to be of interest, the categories in which such justifications might be organized could help both better account for the obligations and better design materials that document interventions.

6.3. Conclusion

Geometry has historically had an important place in the secondary curriculum. Over the years, the arguments for the study of geometry have been diverse (González and Herbst, 2006). Some of those arguments have emphasized the logical organization of the geometry content and the potential that this organization has to teach students about logical reasoning. This mainstream position has stressed the role of clear definitions and axioms as the basis for the development of experiences in which students practice reasoning deductively from premises to conclusion. In this game plan, diagrams have been held somewhat in contempt, as contaminations of pure mathematical reasoning which should see figures as represented by their definitions and theoretical properties only.

Other arguments have come in handy for modifying that logical argument. An intuitive argument that casts geometry as providing a vocabulary for the world of shape and space has also been used to preserve the study of geometry in the curriculum. In this argument, it has been important to educate the capacity for visualization of geometric objects as building blocks of reality. Diagrams and other concrete representations have played an essential role in making this argument possible. Diagrams have been a part of the study of geometry forever, and—as educators we also need to say—fortunately (Dewey, 1903).

A utilitarian argument for the study of geometry has benefited from the intuitive argument's emphasis on naming the real world, but gone beyond it by expecting that knowledge of geometry could assist the needs of intervening to improve the world, particularly measuring and calculating, through the making of artifacts, buildings, and machines. The utilitarian argument hypostatized (or reified) geometry in the real world and helped achieve tremendous accomplishments in architecture and engineering. The recent decades' developments in geographic information systems and in software for designing and prototyping have shown that the interaction between the world of geometric ideas and that of engineering and technology can be very productive for education.

A mathematical argument for the study of geometry emerged in the second half of the twentieth century and recognized the possible role that geometry could play in replicating the activities of doing mathematics. Unlike the logical argument that sought to separate mathematical logic from mathematical substance on the grounds that logical reasoning could be useful in multiple endeavors, the mathematical argument sought to maintain the unity of methodology and focus as it gave students opportunities to experience mathematical discovery. Lakatos' (1976) geometric illustration of the methodological role of informal proof in mathematical discovery has been inspirational for mathematics educators who are interested in students experiencing the use of proof as a tool not only to confirm things but also to find things out.

The modeling approach that we propose in this book takes advantage of all of those arguments. The notion that modeling is a key practice in the doing of mathematics draws legitimacy from the mathematical argument, which is at the center of our approach. The modeling approach casts a productive doubt on the reification proposed by the utilitarian approach. But the impetus to use mathematical modeling to engage with, describe, and control the students' experiences in the world of space and shape is feasible because students come to secondary school with some acquaintance with the language of geometry and because they have cultural and practical needs to make things. The logical argument continues to provide the hope that deductive reasoning skills (and perhaps also the disposition to solve all sorts of problems in one's head) might be transferable and usable in other contexts.

Many questions are unanswered. Can the geometry curriculum be woven continuously as a series of modeling activities? May such a curriculum support students' advancement through the stages of learning trajectories for the various conceptions of figure? May the work of teaching be feasible as teachers try to enact this approach? We hope the readers will take up these ideas and do work that helps us all answer those questions in the coming years.

Notes

1 If a figure is included in another one, the area of the former is smaller or equal to the area of the latter.
2 If a figure is composed of two other figures whose intersection includes no point interior to these other two figures, the area of the first equals the sum of the area of the other two.
3 The lessons were designed around the task of finding a point in a triangle such that when segments from this point to the vertices are drawn, the triangle is divided into three triangles of equal area (see also Herbst, 2006).
4 We use "instruction" to designate the interactions among teacher, students, and content in environments, following the usage of Cohen et al. (2003), and equivalent to expression such as "didactic system" (Chevallard, 1985) or "didactic triangle" (Goodchild and Sriraman, 2012).The alternative sense of instruction as indicating explicitly what someone must do is not intended in our usage.

REFERENCES

Acredolo, L. (1981). Small- and large-scale spatial concepts in infancy and childhood. In L. Liben, A. Patterson, and N. Newcombe (Eds), *Spatial representation and behavior across the life span: Theory and application* (pp. 63–81). New York: Academic Press.

Adler, J. (2005). Mathematics for teaching: What is it and why is it important that we talk about it. *Pythagoras, 62*, 2–11.

Adler, J. and Davis, Z. (2006). Opening another black box: Researching mathematics for teaching in mathematics teacher education. *Journal for Research in Mathematics Education, 37*(4), 270–296.

Aiken, L. R. (1970). Attitudes toward mathematics. *Review of Educational Research, 40*(4), 551–596.

Aiken, L. R. (1976). Update on attitudes and other affective variables in learning mathematics. *Review of Educational Research, 46*(2), 293–31.

Allen, F. B., Douglas, E., Richmond, D., Rickart, C., Swain, H. and Walker, R. (1965). *Geometry. Student's Text.* Part I, Unit 13. Stanford University, CA: School Mathematics Study Group.

Allmendinger, H. (2015). Klein's Elementary Mathematics from a Higher Standpoint – An analysis from a historical and didactic point of view. In K. Bjarnadóttir, F. Furinghetti, J. Prytz and G. Schubring (Eds), *Dig where you stand 3: Proceedings of the Third International Conference on the History of Mathematics Education.* (pp. 9–19). Uppsala, Sweden: Uppsala University.

Anderson. J. R. (1982). Acquisition of cognitive skill. *Psychological Review, 89*(4), 369–406.

Anderson, J. R., Boyle, C. F., Corbett, A. T. and Lewis, M. W. (1990). Cognitive modeling and intelligent tutoring. *Artificial Intelligence, 42*(1), 7–49.

Anderson, J. R., Corbett, A. T., Koedinger, K. R. and Pelletier, R. (1995). Cognitive tutors: Lessons learned. *The Journal of the Learning Sciences, 4*(2), 167–207.

Artigue, M. (1994). Didactical engineering as a framework for the conception of teaching products. In R. Biehler, R. W. Scholz, R. Sträßer and B. Winkelmann (Eds), *Didactics of mathematics as a scientific discipline* (pp. 27–39). Dordrecht, The Netherlands: Kluwer.

Artigue, M., Haspekian, M. and Corblin-Lenfant, A. (2014). Introduction to the Theory of Didactical Situations (TDS). In A. Bikner-Ahsbahs and S. Prediger (Eds), *Networking of theories as a research practice in mathematics education* (pp. 47–65). Cham, Switzerland: Springer.

Artmann, B. (1999). *Euclid book I: Basic geometry. Euclid—the creation of mathematics*. New York: Springer.

Arzarello, F., Micheletti, C., Olivero, F., Robutti, O. and Paola, D. (1998a). A model for analysing the transition to formal proofs in geometry. In A. Olivier and K. Newstead (Eds), *Proceedings of the 22nd International Conference for the Psychology of Mathematics Education*, Vol. 2. (pp. 24–31). Stellenbosch, South Africa: University of Stellenbosch.

Arzarello, F., Micheletti, C., Olivero, F., Robutti, O. and Paola, D. (1998b). Dragging in Cabri and modalities of transition from conjectures to proofs in geometry. In A. Olivier and K. Newstead (Eds), *Proceedings of the 22nd International Conference for the Psychology of Mathematics Education*, Vol. 2. (pp. 32–39). Stellenbosch, South Africa: University of Stellenbosch.

Arzarello, F., Olivero, F., Paola, D. and Robutti, O. (2002). A cognitive analysis of dragging practises in Cabri environments. *Zentralblatt für Didaktik der Mathematik, 34*(3), 66–72.

Azzouni, J. (2013). That we see that some diagrammatic proofs are perfectly rigorous. *Philosophia Mathematica, 21*(3), 323–338.

Baki, A., Kosa, T. and Guven, B. (2011). A comparative study of the effects of using dynamic geometry software and physical manipulatives on the spatial visualisation skills of pre-service mathematics teachers. *British Journal of Educational Technology, 42*(2), 291–310.

Balacheff, N. (1993). La transposition informatique, un nouveau problème pour la didactique. In M. Artigue, R. Gras, C. Laborde and P. Tavignot (Eds), *Vingt ans de didactique des mathématiques en France* (pp. 364–370). Grenoble: La Pensée Sauvage.

Balacheff, N. and Gaudin, N. (2010). Modeling students' conceptions: The case of function. In F. Hitt, D. A. Holton and P. W. Thompson (Eds), *CBMS Issues in Mathematics Education Volume 16: Research in Collegiate Mathematics Education VII* (pp. 207–234). Washington, D.C. American Mathematical Society.

Ball, D. L. (2000). Working on the inside: Using one's own practice as a site for studying teaching and learning. In A. E. Kelly and R. Lesh (Eds), *Handbook of research design in mathematics and science education* (pp. 365–402). Mahwah, NJ: Lawrence Erlbaum.

Ball, D. L. and Bass, H. (2000). Interweaving content and pedagogy in teaching and learning to teach: Knowing and using mathematics. In J. Boaler (Ed.), *Multiple perspectives on the teaching and learning of mathematics* (pp. 83–104). Westport, CT: Ablex Publishing.

Ball, D., Bass, H. and Hill, H. (2004). Knowing and using mathematical knowledge in teaching: Learning what matters. In *Proceedings for the 12th Annual Conference of the South African Association for Research in Mathematics, Science and Technology Education*. Durban: SAARMSTE.

Ball, D. L., Lubienski, S. T. and Mewborn, D. S. (2001). Research on teaching mathematics: The unsolved problem of teachers' mathematical knowledge. In V. Richardson (Ed.), *Handbook of research on teaching*, 4th edition (pp. 433–456). Washington, DC: AERA.

Ball, D. L., Thames, M. H. and Phelps, G. (2008). Content knowledge for teaching what makes it special? *Journal of Teacher Education, 59*(5), 389–407.

Barbin, E. and Menghini, M. (2014). History of teaching geometry. In A. Karp and G. Schubring (Eds), *Handbook on the history of mathematics education* (pp. 473–492). New York, NY: Springer.

Barrantes, M., Blanco, L. J. and Abrantes de Figueredo, C. A. B. (2006). Caracterização das concepções dos professores em formação sobre ensino-aprendizagem da geometria. *Zetetike, 14*(25), 65–92.

Bartholdi, L., Halverscheid, S., Luke, R. and Wardetzky, M. (2011). The collection of mathematical models and instruments: Perspectives for tomorrow. *Georgia Augusta, 8*, 82–83.

Bartocci, C. (2013). "Reasoning well from badly drawn figures": The birth of algebraic topology. *Lettera Matematica, 1*(1–2), 13–22.

Bartolini-Bussi, M. and Mariotti, M. A. (2008). Semiotic mediation in the mathematics classroom: Artifacts and signs after a Vygotskian perspective. In L. English, M. Bartolini-Bussi, G. Jones, R. Lesh and B. Sriraman (Eds), *Handbook of international research in mathematics education*, 2nd edition (pp. 746–783). New York: Routledge.

Bateson, G. (1972). *Steps to an ecology of mind: Collected essays in anthropology, psychiatry, evolution, and epistemology*. Chicago: University of Chicago Press.

Battista, M. T. (2007). The development of geometric and spatial thinking. In F. Lester (Ed.), *Second handbook of research on mathematics teaching and learning* (pp. 843–908). Charlotte, NC: NCTM/Information Age.

Battista, M. T. (2008). Representations and cognitive objects in modern school geometry. In G. Blume and K. Heid (Eds), *Research on technology and the teaching and learning of mathematics (Volume 2): Cases and perspectives* (pp. 341–362). Charlotte, NC: Information Age.

Battista, M. T. (2011). Conceptualizations and issues related to learning progressions, learning trajectories, and levels of sophistication. *The Mathematics Enthusiast, 8*(3), 507–570.

Bauer, M. I. and Johnson-Laird, P. N. (1993). How diagrams can improve reasoning. *Psychological Science, 4*(6), 372–378.

Bautista, A. and Roth, W. M. (2012). The incarnate rhythm of geometrical knowing. *The Journal of Mathematical Behavior, 31*(1), 91–104.

Behrendsen, O. and Götting, E. (1915). *Lehrbuch der Mathematik nach modernen Grundsätzen* (Vol. 2). Leipzig: Buchgesellschaft Teubner.

Ben-Chaim, D., Lappan, G. and Houang, R. T. (1989). Adolescents' ability to communicate spatial information: Analyzing and effecting students' performance. *Educational Studies in Mathematics, 20*(2), 121–146.

Bernard, A., Proust, C. and Ross, M. (2014). Mathematics education in antiquity. In A. Karp and G. Schubring, (Eds) *Handbook on the history of mathematics education* (pp. 27–53). New York: Springer.

Berthelot, R. and Salin, M. H. (1998). The role of pupils' spatial knowledge in the elementary teaching of geometry. In C. Mammana and V.Villani (Eds), *Perspectives on the teaching of geometry for the 21st century: An ICMI study* (pp. 71–77). Dordrecht, The Netherlands: Kluwer.

Biber, Ç., Tuna, A. and Korkmaz, S. (2013). The mistakes and the misconceptions of the eighth grade students on the subject of angles. *European Journal of Science and Mathematics Education, 1*(2), 50.

Bindak, R. (2004). Geometri tutum ölçei güvenirlik çalmas ve bir uygulama (Unpublished doctoral dissertation). Dicle University, Diyarbakir, Turkey.

Birkhoff, G. D. and Beatley, R. (2000). *Basic geometry: Manual for teachers*. Providence, RI: American Mathematical Society. (Original work published 1943).

Bkouche, R. (2009). De l'enseignement de la géométrie. *Reperes—IREM, 76*, 85–103.

Black, M. (1962). *Models and metaphors*. Ithaca, NY: Cornell University Press.

Blanton, M. L. (2002). Using an undergraduate geometry course to challenge pre-service teachers' notions of discourse. *Journal of Mathematics Teacher Education, 5*(2), 117–152.

Blömeke, S. and Delaney, S. (2014). Assessment of teacher knowledge across countries: A review of the state of research. In S. Blömeke, F. J. Hsieh, G. Kaiser and W. Schmidt (Eds), *International perspectives on teacher knowledge, beliefs and opportunities to learn* (pp. 541–585). Dordrecht, The Netherlands: Springer.

Boileau, N. and Herbst, P. (2015, November). Teachers' Expectation About Geometric Calculations in High School Geometry. In Bartell, T. G., Bieda, K. N., Putnam, R. T., Bradfield, K., & Dominguez, H. (Eds.). *Proceedings of the 37th annual meeting of the North American Chapter of the International Group for the Psychology of Mathematics Education* (pp. 269–276), East Lansing, MI: Michigan State University.

Bonola, R. (1955). *Non-Euclidean geometry: A critical and historical study of its development*. New York: Dover.

Bourbaki, N. (1950). The architecture of mathematics. *American Mathematical Monthly, 57*(4), 221–232.

Bourdieu, P. (1991). *The logic of practice*. Stanford, CA: Stanford University Press.

Bos, H. J. (1993). 'The bond with reality is cut'—Freudenthal on the foundations of geometry around 1900. *Educational Studies in Mathematics, 25*(1–2), 51–58.

Boyd, A. and Rubin, A. (1996). Interactive video: A bridge between motion and math. *International Journal of Computers for Mathematical Learning, 1*(1), 57–93.

Bright, G. W., Waxman, H. C. and Williams, S. E. (Eds). (1994). *Impact of calculators on mathematics instruction*. Lanham, MD: University Press of America.

Brock, W. H. (1975). Geometry and the universities: Euclid and his modern rivals 1860–1901. *History of Education, 4*(2), 21–35.

Brousseau, G. (1997). *Theory of didactical situations in mathematics: Didactique des Mathematiques 1970–1990*. Dordrecht, The Netherlands: Kluwer.

Bruce, C. D. and Hawes, Z. (2015). The role of 2D and 3D mental rotation in mathematics for young children: what is it? Why does it matter? And what can we do about it? *ZDM The International Journal of Mathematics Education, 47*(3), 331–343.

Brunner-Traut, E. (1974). Epilogue: Aspective. In H. Schäfer, *Principles of Egyptian art* (Edited with an epilogue by E. Brunner-Traut; translated and edited with an introduction by J. Baines). Oxford: Clarendon Press (Original work published 1919).

Burger, W. F. and Shaughnessy, J. M. (1986). Characterizing the van Hiele levels of development in geometry. *Journal for Research in Mathematics Education, 17*(1), 31–48.

Cabassut, R. (2009). The double transposition in proving. In F. L. Lin, F. J. Hsieh, G. Hanna, and M. de Villiers (Eds.), *Proceedings of the ICMI Study 19 conference: Proof and proving in mathematics education* (Vol. 1, pp. 112–117). Taipei: National Taiwan Normal University.

Cai Shanghe 蔡上鶴. (2002). Xin zhongguo zhongxue shuxue jiaocai jianshe wushiyi nian 新中国中学数学教材建 51年 (Fifty-one years of mathematical textbooks in New China). *Shuxue tongbao* 数学通报 *Bulletin des Sciences Mathematics, 9*, 14–18.

Cantürk-Gunhan, B. and Baser, N. (2007). The development of self-efficacy scale toward geometry. *Hacettepe Universitesi Egitim Fakultesi dergisi-Hacettepe University Journal of Education, 33*, 68–76.

Carpenter, T. P., Fennema, E., Peterson, P. L., Chiang, C. P. and Loef, M. (1989). Using knowledge of children's mathematics thinking in classroom teaching: An experimental study. *American Educational Research Journal, 26*(4), 499–531.

Carter, G. and Norwood, K. S. (1997). The relationship between teacher and student beliefs about mathematics. *School Science and Mathematics, 97*(2), 62–67.

Catalã, J. S. (1994). *Ciencia y técnica en la metropolización de América*. Madrid: Theatrum Machinae.

Chazan, D. (1993). High school geometry students' justification for their views of empirical evidence and mathematical proof. *Educational Studies in Mathematics, 24*(4), 359–387.

Chazan, D. (2000). *Beyond Formulas in Mathematics and Teaching: Dynamics of the High School Algebra Classroom*. New York: Teachers College Press.

Chazan, D., Herbst, P. and Clark, L. (2016). Research on the teaching of mathematics: A call to theorize the role of society and schooling in mathematics. In D. Gitomer and C. Bell (Eds), *Handbook of research on teaching*, 5th edition (pp. 1039–1097). Washington, DC: AERA.

Chazan, D. and Sandow, D. (2011). Why did you do that? Reasoning in algebra classrooms. *The Mathematics Teacher, 104*(6), 460–464.

Chazan, D. and Yerushalmy, M. (1998). Charting a course for secondary geometry. In R. Lehrer and D. Chazan (Eds), *Designing learning environments for developing understanding of geometry and space* (pp. 67–90). Mahwah, NJ: Erlbaum.

Chen, C. L. and Herbst, P. (2013). The interplay among gestures, discourse, and diagrams in students' geometrical reasoning. *Educational Studies in Mathematics, 83*(2), 285–307.

Cheng, P. W., Holyoak, K. J., Nisbett, R. E. and Oliver, L. M. (1986). Pragmatic versus syntactic approaches to training deductive reasoning, *Cognitive Psychology, 18*(3), 293–328.

Chevallard, Y. (1985). *La transposition didactique: Du savoir savant au savoir enseignée.* Grenoble: La Pensée Sauvage.

Chinnappan, M. and Lawson, M.J. (2005). A framework for analysis of teachers' geometric content knowledge and geometric knowledge for teaching. *Journal of Mathematics Teacher Education, 8*(3), 197–221.

Chinnappan, M., Nason, R. and Lawson, M. (1996). Pre-service teachers' pedagogical and content knowledge about trigonometry and geometry: An initial investigation. In P. Clarkson (Ed.), *Proceedings of the 19th Annual Conference of the Mathematics Education Research Group of Australasia* (pp. 115–122). Melbourne, MERGA.

Christou, C., Jones, K., Mousoulides, N. and Pittalis, M. (2006). Developing the 3DMath dynamic geometry software: theoretical perspectives on design. *International Journal for Technology in Mathematics Education, 13*(4), 168–174.

Cirillo, M. (2013). Mathematical proof tools: Supporting the introduction to formal proof. In M. Martinez and A. Castro Superfine (Eds.). *Proceedings of the 35th annual meeting of the North American chapter of the International Group for the Psychology of Mathematics Education* (pp. 203–206). Chicago, IL: University of Illinois at Chicago.

Clements, D.H. (2004). Geometric and spatial thinking in early childhood education. In D. Clements, J. Sarama and M. A. DiBaise (Eds), *Engaging young children in mathematics: Results of the conference on standards for pre-school and kindergarten mathematics education* (pp. 267–298). Mahwah, NJ: Erlbaum.

Clements, D. H. and Battista, M. T. (1992). Geometry and spatial reasoning. In D. A. Grouws (Ed.) *Handbook of research on mathematics teaching and Learning* (pp. 420–464). New York: NCTM/Macmillan.

Clements, D. H., Battista, M. T. and Sarama, J. (2001). Logo and geometry. *Journal for Research in Mathematics Education Monograph Series*, 10.

Clements, D. H., Battista, M. T., Sarama, J. and Swaminathan, S. (1996). Development of turn and turn measurement concepts in a computer-based instructional unit. *Educational Studies in Mathematics, 30*(4), 313–337.

Clements, D. H., Battista, M. T., Sarama, J., Swaminathan, S. and McMillen, S. (1997). Students' development of length concepts in a logo-based unit on geometric paths. *Journal for Research in Mathematics Education, 20*(1), 70–95.

Clements, D. and Sarama, J. (2014). *Learning and Teaching Early Math: The Learning Trajectories Approach* (2nd edition). New York, NY: Routledge.

Clements, D. H., Swaminathan, S., Hannibal, M. A. Z. and Sarama, J. (1999). Young children's concepts of shape. *Journal for Research in Mathematics Education, 30*(2), 192–212.

Clements, D. H., Wilson, D. C. and Sarama, J. (2004). Young children's composition of geometric figures: A learning trajectory. *Mathematical Thinking and Learning, 6*(2), 163–184.

Clements, M. A. and Ellerton, N. F. (1996). *Mathematics education research: Past, present and future.* Bangkok, Thailand: UNESCO.

Cobb, P. (2000). Conducting teaching experiments in collaboration with teachers. In A. E. Kelly and R. Lesh (Eds), *Handbook of research design in mathematics and science education* (pp. 307–333). Mahwah, NJ: Lawrence Erlbaum.

Cobb, P., Confrey, J., diSessa, A., Lehrer, R. and Schauble, L. (2003). Design experiments in educational research. *Educational Researcher, 32*(1), 9–13.

Cohen, D. K. (2005). Professions of human improvement: Predicaments of teaching. In M. Nisan and O. Schremer (Eds.). *Educational deliberations* (pp. 278–294). Jerusalem: Keter Publishers.

Cohen, D. K., Raudenbush, S. W. and Ball, D. L. (2003). Resources, instruction, and research. *Educational Evaluation and Policy Analysis, 25*(2), 119–142.

Coliva, A. (2012). Human diagrammatic reasoning and seeing-as. *Synthese, 186*(1), 121–148.

Conference Board for the Mathematical Sciences (2000). *Mathematical education of teachers.* Washington, DC: CBMS.

Confrey, J. and Stohl, V. (Eds). (2004). *On evaluating curricular effectiveness: Judging the quality of K-12 mathematics evaluations.* Washington, DC: National Academies Press.

Conway, J., Doyle, P., Gilman, J. and Thurston, W. (1991). *Geometry and the imagination.* Retrieved from http://www.geom.uiuc.edu/docs/education/institute91/handouts/handouts.html

Cooper, B. (1985). *Renegotiating secondary school mathematics: A study of curriculum change and stability.* London: Taylor & Francis.

Cornelius, M. L. (1973). Euclid must go? *Mathematics in School 2*(2), 16–17.

Corry, L. (1996). *Modern algebra and the rise of mathematical structures.* Basel: Birkhauser.

Costanzo, J. M. (2008). The Euclidean mousetrap: Schopenhauer's criticism of the synthetic method in geometry, *Idealistic Studies, 38*(3), 209–220.

Cuban, L. (1992). Curriculum stability and change. In P. W. Jackson (Ed.), *Handbook of research on curriculum,* (pp. 216–247). New York: Macmillan.

Curriculum Planning and Development Division (2006). *Secondary Mathematics Syllabuses.* Singapore: Ministry of Education.

D'Ambrosio, U., Dauben, J. and Parshall, K. H. (2014). Mathematics education in America in the premodern period. In A. Karp and G. Schubring (Eds), *Handbook on the history of mathematics education* (pp. 175–196). New York: Springer.

Damerow, P. and Westbury, I. (1985). Mathematics for all: Problems and implications. *Journal of Curriculum Studies, 17*(2), 175–184.

Dauben, J. W., Yee, L. P., Raina, D. and Xu, Y. (2014). Mathematics education in modern Asia. In A. Karp and G. Schubring (Eds), *Handbook on the history of mathematics education* (pp. 361–390). New York: Springer.

Davidson, J. (1832). *A system of practical mathematics.* Bell and Bradfute.

Depaepe, F., Verschaffel, L. and Kelchtermans, G. (2013). Pedagogical content knowledge: A systematic review of the way in which the concept has pervaded mathematics educational research. *Teaching and Teacher Education, 34,* 12–25.

Devichi, C. and Munier, V. (2013). About the concept of angle in elementary school: Misconceptions and teaching sequences. *Journal of Mathematical Behavior, 32,* 1–19.

de Carvalho, J.B.P. (2014). Mathematics education in Latin America. In A. Karp and G. Schubring (Eds), *Handbook on the history of mathematics education* (pp. 335–360). New York: Springer.

de Villiers, M. (1990). The role and function of proof in mathematics. *Pythagoras, 24*(1), 17–24.

de Villiers, M. (1994). The role and function of hierarchical classification of quadrilaterals. *For the Learning of Mathematics, 14*(1), 11–18.

de Villiers, M. (1998). To teach definitions in geometry or teach to define?, In A. Olivier and K. Newstead (Eds), *Proceedings of the 22nd International Conference for the Psychology of Mathematics Education,* Vol. 2. (pp. 248–255). Stellenbosch, South Africa: University of Stellenbosch

de Villiers, M. (2003). *Rethinking proof with The Geometer's Sketchpad.* San Francisco, CA: Key Curriculum Press.

de Villiers, M. (2004). Using dynamic geometry to expand mathematics teachers' understanding of proof. *International Journal of Mathematical Education in Science and Technology, 35*(5), 703–724.

Dewey, J. (1903). The psychological and the logical in teaching geometry. *Educational Review, 25,* 387–399.

Dewey, J. (1933). *How we think*. Boston, MA: D.C. Heath.

Dicks, D. R. (1959). Thales. *The Classical Quarterly* (New Series), *9*(3–4), 294–309.

Dienes, Z. P. (1969). *Building Up Mathematics*. (Revised Edition) London: Hutchinson Educational.

Dieudonné, J. (1959). *New thinking in school mathematics*. Asnière-sur-Oise: The Royaumont Seminar.

Dimmel, J. K. and Herbst, P. G. (2015a). The semiotic structure of geometry diagrams: How textbook diagrams convey meaning. *Journal for Research in Mathematics Education*, *46*(2), 147–195.

Dimmel, J. K. and Herbst, P. G. (2015b). Investigating secondary mathematics teachers' attitudes toward alternative communication practices while doing proofs in geometry. In T. G. Bartell, K. N. Bieda, R. T. Putnam, K. Bradfield, and H. Dominguez (Eds.). *Proceedings of the 37th annual meeting of the North American Chapter of the International Group for the Psychology of Mathematics Education* (pp. 277–284), East Lansing, MI: Michigan State University.

Dodgson, C. L. (1879). *Euclid and his modern rivals*. London: Macmillan.

Doyle, W. (1988). Work in mathematics classes: The context of students' thinking during instruction. *Educational Psychologist*, *23*(2), 167–180.

Dreyfus, T. (1995). Imagery for diagrams. In R. Sutherland and J. Mason (Eds), *Exploiting mental imagery with computers in mathematics education* (pp. 3–19). Berlin: Springer.

Drijvers, P. and Doorman, M. (1996). The graphics calculator in mathematics education. *The Journal of Mathematical Behavior*, *15*(4), 425–440.

Duval, R. (1995). Geometrical pictures: Kinds of representation and specific processings. In R. Sutherland and J. Mason (Eds), *Exploiting mental imagery with computers in mathematics education* (pp. 142–157). Berlin: Springer.

Duval, R. (1998). Geometry from a cognitive point of view. In C. Mammana and V. Villani, (Eds), *Perspectives on the teaching of geometry for the 21st century* (pp. 37–51). Dordrecht, The Netherlands: Kluwer.

Duval, R. (2002). Proof understanding in mathematics: What ways for students? In F. L. Lin (Ed.), *Understanding proving and proving to understand* (pp. 61–77). Taipei: National Taiwan Normal University.

Duval, R. (2005). Les conditions cognitives de l'apprentissage de la géométrie: développement de la visualisation, différenciation des raisonnements et coordination de leurs fonctionnements. *Annales de Didactique et de Sciences Cognitives*, *10*, 5–53.

Eichler, A. and Erens, R. (2015). Domain-specific belief systems of secondary mathematics teachers. In B. Pepin and B. Rosken-Winter (Eds), *From beliefs to dynamic affect systems in mathematics education* (pp. 179–200). Cham, Switzerland: Springer International Publishing.

Eliot, C. (1905). The fundamental assumptions in the report of the Committee of Ten (1893). *Educational Review*, *30*, 325–343.

Eliot, C., Harris, W., Angell, J., Tetlow, J., Taylor, J., Robinson, O., Baker, J., Jesse, R. McKenzie, J. and King, H. (1969). Report of the Committee of Ten to the National Education Association. In National Education Association, *Report of the Committee on secondary school studies*, Arno Press, New York, pp. 3–5. (Original work published 1893.).

Emch, A. (1914). Goethe and Schopenhauer on mathematics. *The Open Court*, *1914*(9), 2.

Euclid (1956). *The thirteen books of Euclid's elements*, translated by Thomas L. Heath, 2nd edition. New York: Dover Publications.

Eves, H. W. (1990). *An introduction to the history of mathematics*. Philadelphia, PA: Saunders.

Fehr, H. F. (1963). Reform of instruction in geometry. *The American Mathematical Monthly*, *70*(3), 323–327.

Fennema, E. and Franke, M. L. (1992). Teachers' knowledge and its impact. In D. Grouws (Ed.), *Handbook of research on mathematics teaching and learning* (pp. 147–164). New York: Macmillan.

Fennema, E. and Sherman, J. A. (1976). Fennema-Sherman mathematics attitudes scales: Instruments designed to measure attitudes toward the learning of mathematics by females and males. *Journal for Research in Mathematics Education*, 7(5), 324–326.

Ferrara, F. and Mammana, M.F. (2013). Close your eyes and see. . . An approach to spatial geometry. In B. Ubuz, C. Haser, and M. A. Mariotti (Eds), *Proceedings of the Eighth Congress of the European Society for Research in Mathematics Education* (pp. 625–634). Ankara, Turkey: Middle East Technical University.

Ferrini-Mundy, J. and Findell, B. (2001). The mathematical education of prospective teachers of secondary school mathematics: Old assumptions, new challenges. *CUPM discussion papers about mathematics and the mathematical sciences in 2010: What should students know*, 31–41. Washington, DC: Mathematical Association of America.

Fey, J.T. (Ed.) (1984). *Computing and mathematics. The impact on secondary school curricula. Report of a conference sponsored by the National Science Foundation* (College Park, Maryland, October, 1982). Reston, VA: National Council of Teachers of Mathematics.

Fischbein, E. (1993). The theory of figural concepts. *Educational Studies in Mathematics*, *24*(2), 139–162.

Fisher, K.R., Hirsh-Pasek, K., Newcombe, N. and Golinkoff, R.M. (2013). Taking shape: Supporting preschoolers' acquisition of geometric knowledge through guided play. *Child Development*, *84*(6), 1872–1878.

Foletta, G. (1994). Technology and guided inquiry: Understanding of students' thinking while using a cognitive computer tool, the Geometer's Sketchpad, in a geometry class. Unpublished doctoral dissertation. University of Iowa, Iowa City, IA.

Forbes, J.E. (1973). An eclectic program in geometry. In K. B. Henderson (Ed.) (1973). *Geometry in the Mathematics Curriculum* (pp. 334–366). Reston, VA: NCTM.

Forsythe, S.K. (2015). Dragging maintaining symmetry: can it generate the concept of inclusivity as well as a family of shapes? *Research in Mathematics Education*, *17*(3), 198–219.

Franke, M.L., Kazemi, E. and Battey, D. (2007). Mathematics teaching and classroom practice. In F. Lester (Ed.), *Second handbook of research on mathematics teaching and learning*, 1, 225–256. Charlotte,NC: Information Age.

Frege, G. (1980). *Gottlob Freges Briefwechsel mit D. Hilbert, E. Husserl, B. Russell sowie ausgewählte Einzelbriefe Freges* (Vol. 321). Hamburg: Meiner Verlag.

Freudenthal, H. (1957). Zur Geschichte der Grundlagen der Geometrie - zugleich eine Besprechung der 8. Aufl. von Hilberts Grundlagen der geometrie. *Nieuw Archiefvoor Wiskunde* *3*(5), 105–142.

Freudenthal, H. (1966). The main trends in the foundations of geometry in the 19th century. *Studies in Logic and the Foundations of Mathematics*, *44*, 613–621.

Freudenthal, H. (1971). Geometry between the devil and the deep sea. In: Steiner, H. G. (Ed.), *The teaching of geometry at the pre-college level* (pp. 137–159). Dordrecht, The Netherlands: Reidel.

Freudenthal, H. (1973). *Mathematics as an educational task*. Dordrecht, The Netherlands: Reidel.

Fried, M. (2014). Mathematics and mathematics education: Beginning a dialogue in an atmosphere of increasing estrangement. In M. N. Fried and T. Dreyfus (Eds), *Mathematics & mathematics education: Searching for common ground* (pp. 25–33). Dordrecht, The Netherlands: Springer.

Frostig, M. (1963). Visual perception in the brain-injured child. *American Journal of Orthopsychiatry*, *33*(4), 665–671.

Fujii, T. and Matano, H. (Eds). (2012). *Mathematics international grade 8*. Tokyo: Tokyo Shoseki.

Fujita, T. (2001). The order of theorems in the teaching od Euclidean geometry: Learning from developments in textbooks in the early 20th Century. *Zentralblatt für Didaktik der Mathematik, 33*(6), 196–203.

Fujita, T. (2012). Learners' level of understanding of the inclusion relations of quadrilaterals and prototype phenomenon. *The Journal of Mathematical Behavior, 31*(1), 60–72.

Fujita, T. and Jones, K. (2003). The place of experimental tasks in geometry teaching: learning from the textbook designs of the early 20th century. *Research in Mathematics Education, 5*(1), 47.

Fujita, T. and Jones, K. (2014). Reasoning-and-proving in geometry in school mathematics textbooks in Japan. *International Journal of Educational Research, 64*, 81–91.

Fuys, D., Geddes, D. and Tischler, R. (1984). *English Translation of Selected Writings of Dina van Hiele-Geldof and Pierre M. van Hiele. ERIC Document 287697.* New York: Brooklyn College.

Fuys, D., Geddes, D. and Tischler, R. (1988). The van Hiele model of thinking in geometry among adolescents. *Journal for Research in Mathematics Education. Monograph* 3,

Fyhn, A.B. (2008). A climbing class' reinvention of angles. *Educational Studies in Mathematics, 67*(1), 19–35

Gal, H. and Linchevski, L. (2010). To see or not to see: Analyzing difficulties in geometry from the perspective of visual perception. *Educational Studies in Mathematics, 74*(2), 163–183.

Galuzzi, M. and Neubrand, M. (1998). Influences of history and the general social development on the teaching of geometry. In C. Mammana and V. Villani, (Eds.), *Perspectives on the teaching of geometry for the 21st century: An ICMI study* (pp. 229–234). Dordrecht, The Netherlands: Kluwer.

Garfinkel, H. and Sacks, H. (1970). On formal structures of practical actions. In J. C. McKinney and E. A. Tiryakian (Eds.), *Theoretical sociology* (pp. 337–366). New York: Appleton-Century Crofts.

Giacardi, L. (2006). From Euclid as textbook to the Giovanni Gentile reform (1867–1923): Problems, methods and debates in mathematics teaching in Italy. *Paedagogica Historica*, 42(4–5), 587–613.

Giacardi, L. and Scoth, R. (2014). Secondary school mathematics teaching from the early nineteenth century to the mid-twentieth century in Italy. In A. Karp and G. Schubring (Eds), *Handbook on the history of mathematics education*, (pp. 201–228). New York: Springer.

Ginsberg, H. and Opper, S. (1969). *Piaget's theory of intellectual development*. Englewood Cliffs, NJ: Prentice-Hall.

Girnat, B. (2011). Ontological beliefs and their impact on teaching elementary geometry. *PNA, 5*(2), 37–48.

Gispert, H. (2014). Mathematics education in France: 1800–1900. History of teaching geometry. In *Handbook on the history of mathematics education* (pp. 473–492). New York: Springer.

Glas, E. (2002). Socially conditioned mathematical change: the case of the French revolution. *Studies in History and Philosophy of Science Part A, 33*(4), 709–728.

Glymour, C. (1977). The epistemology of geometry. *Noûs, 11*, 227–251.

Godfrey, C. (1910). The board of education circular on the teaching of geometry. *The Mathematical Gazette, 5*(84), 195–200.

Godfrey, C. and Siddons, A.W. (1931). *The teaching of elementary mathematics*. Cambridge: Cambridge University Press.

Goldenberg, E. P. and Cuoco, A. A. (1998). What is dynamic geometry. In Lehrer, R. and Chazan, D. (Eds), *Designing learning environments for developing understanding of geometry and space* (351–368). Mahwah, NJ: Erlbaum.

Gonzalez, G. and Herbst, P. G. (2006). Competing arguments for the geometry course: Why were American high school students supposed to study geometry in the twentieth century? *International Journal for the History of Mathematics Education, 1*(1), 7–33.

González, G. and Herbst, P. G. (2009). Students' conceptions of congruency through the use of dynamic geometry software. *International Journal of Computers for Mathematical Learning, 14*(2), 153–182.

Gonzato, M., Fernández Blanco, M. T. and Godino, J. D. (2011). Tareas para el desarrollo de habilidades de visualización y orientación espacial. *Números: Revista de Didáctica de las Matemáticas*, 77, 99–117.

Gonzato, M. and Godino, J. D. (2010). Aspectos históricos, sociales y educativos de la orientación espacial. *Revista Iberoamericana de Educación Matemática, 23*, 45–58.

Goodchild, S. and Sriraman, B. (2012). Revisiting the didactic triangle: from the particular to the general. *ZDM The International Journal on Mathematics Education, 44*(5), 581–585.

Gooya, Z. (2007). Mathematics teachers' beliefs about a new reform in high school geometry in Iran. *Educational Studies in Mathematics, 65*(3), 331–347.

Grattan-Guinness, I. (1998). *The Norton History of the Mathematical Sciences: The Rainbow of Mathematics*. New York: Norton.

Gravemeijer, K. (1990). Realistic geometry education. *Research in Mathematics Education, 11*, 79–91.

Greenberg, M. J. (2010). Old and new results in the foundations of elementary plane euclidean and Non-Euclidean geometries. *The American Mathematical Monthly, 117*(3), 198–219.

Greenstein, S. (2014). Making sense of qualitative geometry: The case of Amanda. *The Journal of Mathematical Behavior, 36*, 73–94.

Greenstreet, W. J. (1894). Herbart's view of the place of mathematics in education. *The Mathematical Gazette, 2*, 9–11.

Griffiths, H. B. (1998). The British experience. In Mammana, C. and Villani, V. (Eds). *Perspectives on the Teaching of Geometry for the 21st Century: An ICMI Study* (pp.194–204). Dordrecht, The Netherlands: Kluwer.

Grigutsch, S., Raatz, U. and Törner, G. (1998). Einstellungen gegenüber Mathematik bei Mathematiklehrern. *Journal für Mathematik-Didaktik, 19*(1), 3–45

Grossman, P. L. (1990). *The making of a teacher: Teacher knowledge and teacher education*. New York: Teachers College Press.

Grover, B. W. and Connor, J. (2000). Characteristics of the college geometry course for preservice secondary teachers. *Journal of Mathematics Teacher Education, 3*(1), 47–67.

Guggenheimer, H. (1967). *Plane geometry and its groups*. San Francisco: Holden Day.

Gutiérrez, A., Jaime, A. and Fortuny, J. M. (1991). An alternative paradigm to evaluate the acquisition of the van Hiele levels. *Journal for Research in Mathematics Education, 22*(3), 237–251.

Gutzmer, A. (1905). Bericht betreffend den Unterricht in der Mathematik an den neunklassigen höheren Lehranstalten. *Zeitschrift für mathematischen und naturwissenschaftlichen Unterricht, 36*, 543–553.

Gutzmer, A. (Ed.) (1908). *Die Tätigkeit der Unterrichtskommission der Gesellschaft deutscher Naturforscher und Ärzte*. Leipzig: Teuben.

Hadas, N., Hershkowitz, R. and Schwarz, B.B. (2000). The role of contradiction and uncertainty in promoting the need to prove in dynamic geometry environments. *Educational Studies in Mathematics, 44*(1–2), 127–150.

Halat, E. (2008). Pre-service elementary school and secondary mathematics teachers' Van Hiele levels and gender differences. *Issues in the Undergraduate Mathematics Preparation of School Teachers*, 1.

Halsted, G. (1893). The old and the new geometry. *Educational Review, 6*, 144–157.

Hamlin, M.L. (2005). Lessons in Educational Equity: Opportunities for Learning in an Informal Geometry Class. Unpublished doctoral dissertation. University of Michigan, Ann Arbor.

Handal, B., Campbell, C., Cavanagh, M., Petocz, P. and Kelly, N. (2013). Technological pedagogical content knowledge of secondary mathematics teachers. *Contemporary Issues in Technology and Teacher Education, 13*(1), 22–40.

Hanna, G. (1983). *Rigorous proof in mathematics education.* Toronto, ON: Ontario Institute for Studies.

Hanna, G. (2000). Proof, explanation and exploration: An overview. *Educational Studies in Mathematics, 44*(1–2), 5–23.

Hanna, G. and de Bruyn, Y. (1999). Opportunity to learn proof in Ontario grade twelve mathematics texts. *Ontario Mathematics Gazette, 37*(4), 23–29.

Hannafin, R. D., Burruss, J. D. and Little, C. (2001). Learning with dynamic geometry programs: Perspectives of teachers and learners. *Journal of Educational Research, 94,* 132–144.

Hansen, V.L. (1998). Everlasting geometry. In C. Mammana and V. Villani (Eds.), *Perspectives on the teaching of geometry for the 21st century: An ICMI study* (pp. 1–8). Dordrecht, The Netherlands: Kluwer.

Harel, G. and Sowder, L. (1998). Students' proof schemes: Results from exploratory studies. *Research in collegiate mathematics education III,* 7, 234–282.

Hart, L.C. (2002). Preservice teachers' beliefs and practice after participating in an integrated content/methods course. *School Science and Mathematics, 102*(1), 4–14.

Hartshorne, R. (2000a). *Geometry: Euclid and beyond.* New York: Springer.

Hartshorne, R. (2000b). Teaching geometry according to Euclid. *Notices of the AMS, 47*(4), 460–465.

Hattie, J. (2003). *Teachers make a difference: What is the research evidence?* Melbourne: Australian Council for Educational Research.

Hattie, J. (2009). *Visible learning: A synthesis of over 800 meta-analyses relating to achievement.* London: Routledge.

Healy, L., Hölzl, R., Hoyles, C. and Noss, R. (1994). Messing up. *Micromath, 10,* 14–17.

Hefendehl-Hebeker, L. (2000). *Figuren und Abbildungen im Geometrieunterricht der Sekundarstufe I.* Augsburg: Wißner-Verlag.

Henderson, K.B. (Ed.) (1973). *Geometry in the mathematics curriculum.* Reston, VA: NCTM.

Hendriks, M., Ippersiel, M., Kortenkamp, U., Kreis, Y., Laborde, C., Pech, P., Recio, T. and Wasserman, A. (2008). *Status quo report on DGS usage.* ECP 2006 EDU 410016 Intergeo. Retrieved from http://i2geo.net/files/i2g_status_quo_report_jan2008.pdf on June 14, 2015.

Henningsen, M. and Stein, M.K. (1997). Mathematical tasks and student cognition: Classroom-based factors that support and inhibit high-level mathematical thinking and reasoning. *Journal for Research in Mathematics Education, 28*(5), 524–549.

Herbst, P. G. (2002a). Establishing a custom of proving in American school geometry: Evolution of the two-column proof in the early twentieth century. *Educational Studies in Mathematics, 49*(3), 283–312.

Herbst, P. (2002b). Engaging students in proving: A double bind on the teacher. *Journal for Research in Mathematics Education, 33,* 176–203.

Herbst, P. (2003). Using novel tasks to teach mathematics: Three tensions affecting the work of the teacher. *American Educational Research Journal, 40*(1), 197–238.

Herbst, P. (2004). Interactions with diagrams and the making of reasoned conjectures in geometry. *Zentralblatt für Didaktik der Mathematik, 36*(5), 129–139.

Herbst, P. (2005). Knowing about "equal area" while proving a claim about equal areas. *Recherches en Didactique des Mathématiques, 25*(1), 11–56.

Herbst, P. (2006). Teaching geometry with problems: Negotiating instructional situations and mathematical tasks. *Journal for Research in Mathematics Education, 37*(4), 313–347.

Herbst, P. (2008). The teacher and the task. In O. Figueras, J. Cortina, S. Alatorre, T. Rojano, and A. Sepúlveda (Eds.), *Proceedings of the 32th Annual Meeting of the International*

Group for the Psychology of Mathematics Education. (Vol. 1, 125–131). Morelia, Mexico: Universidad Michoacana de San Nicolás de Hidalgo.

Herbst, P. (2010). Practical rationality and the justification for actions in mathematics teaching. In P. Brosnan, D. Erchick, and L. Flevares (Eds.), *Proceedings of the 32nd North American Chapter of the International Group for the Psychology of Mathematics Education*. Columbus, OH: The Ohio State University.

Herbst, P. and Balacheff, N. (2009). Proving and knowing in public: What counts as proof in a classroom. In M. Blanton, D. Stylianou, and E. Knuth (Eds.), *Teaching and learning proofs across the grades. A K-16 perspective*. (pp. 40–63). New York: Routledge.

Herbst, P. and Brach, C. (2006). Proving and doing proofs in high school geometry classes: What is it that is going on for students? *Cognition and Instruction*, *24*(1), 73–122.

Herbst, P. and Chazan, D. (2003). Exploring the practical rationality of mathematics teaching through conversations about videotaped episodes: The case of engaging students in proving. *For the Learning of Mathematics*, *23*(1), 2–14.

Herbst, P. and Chazan, D. (2011). Research on Practical Rationality: Studying the Justification of Actions in Mathematics Teaching. *The Mathematics Enthusiast*, 8(3), 405–462.

Herbst, P. and Chazan, D. (2012). On the instructional triangle and sources of justification for actions in mathematics teaching. *ZDM The International Journal on Mathematics Education*, *44*(5), 601–612.

Herbst, P. and Chazan, D. (in press). The role of theory development in increasing the subject specificity of research on mathematics teaching. In J. Cai (Ed.), *First Compendium for Research in Mathematics Education*. Reston, VA: NCTM.

Herbst, P. Chen, C., Weiss, M. and González, G., with Nachlieli, T., Hamlin, M. and Brach, C. (2009). "Doing proofs" in geometry classrooms. In M. Blanton, D. Stylianou, and E. Knuth (Eds), *Teaching and learning of proof across the grades: A K-16 perspective* (pp. 250–268). New York: Routledge.

Herbst, P. and Dimmel, J. (2011, April). *Teaching geometry through problems and its demands of knowledge management*. Paper presented at the 2011 Annual Meeting of the American Educational Research Association. New Orleans, LA.

Herbst, P. with González, G., Hsu, H. Y., Chen, C., Weiss, M. and Hamlin, M. (2010). Instructional situations and students' opportunities to reason in the high school geometry class. Manuscript. *Deep Blue at the University of Michigan*. http://hdl.handle.net/2027.42/78372

Herbst, P., Gonzalez, G. and Macke, M. (2005). How can geometry students understand what it means to "define" in mathematics? *The Mathematics Educator*, *15*(2), 17–24.

Herbst, P. and Kosko, K. W. (2012). Mathematical knowledge for teaching high school geometry. In L. R. Van Zoest, J. J. Lo and J. L. Kratky (Eds), *Proceedings of the 34th annual meeting of the North American Chapter of the International Group for the Psychology of Mathematics Education* (pp. 438–444), Kalamazoo, MI: Western Michigan University.

Herbst, P. and Kosko, K. W. (2014). Mathematical knowledge for teaching and its specificity to high school geometry instruction. In J. Lo, K. R. Leatham, and L. R. Van Zoest (Eds), *Research trends in mathematics teacher education* (pp. 23–46). New York: Springer.

Herbst, P., Kosko, K. and Dimmel, J. (2013). How are geometric proof problems presented? Conceptualizing and measuring teachers' recognition of the diagrammatic register. In M. Martinez and A. Castro Superfine (Eds), *Proceedings of the 35th annual meeting of the North American chapter of the International Group for the Psychology of Mathematics Education* (pp. 179–186). Chicago, IL: University of Illinois at Chicago.

Herbst, P. and Miyakawa, T. (2008). When, how, and why prove theorems? A methodology for studying the perspective of geometry teachers. *Zentralblatt für Didaktik der Mathematik*, *40*(3), 469–486.

Herbst, P., Nachlieli, T. and Chazan, D. (2011). Studying the practical rationality of mathematics teaching: What goes into "installing" a theorem in geometry?. *Cognition and Instruction, 29*(2), 218–255.

Hershkowitz, R. (1990), Psychological aspects of learning geometry. In P. Nesher and J. Kilpatrick (Eds) *Mathematics and Cognition* (pp. 70–95). Cambridge: Cambridge University Press.

Hiebert, J., Gallimore, R., Garnier, H., Givvin, K. B., Hollingsworth, H., Jacobs, J., Chui, A. M-Y., Wearne, D., Smith, M., Kersting, N., Manaster, A., Tseng, E., Etterbeek, W., Manaster, C., Gonzales, P. and Stigler, J. W. (2003). *Teaching mathematics in seven countries: Results from the TIMSS 1999 Video Study* (NCES 2003–013). Washington, DC: U.S. Department of Education, National Center for Education Statistics.

Hiebert, J. and Grouws, D. A. (2007). The effects of classroom mathematics teaching on students' learning. In F. Lester (Ed.), *Second handbook of research on mathematics teaching and learning*, 1, 371–404. Charlotte, NC: Information Age.

Hilbert, D. (1902). *The foundations of geometry*. Peru, IL: Open Court.

Hilbert, D. and Cohn-Vossen, S. (1999). *Geometry and the Imagination*. Providence, RI: American Mathematical Society.

Hill, H. C., Blunk, M. L., Charalambous, C. Y., Lewis, J. M., Phelps, G. C., Sleep, L. and Ball, D.L. (2008). Mathematical knowledge for teaching and the mathematical quality of instruction: An exploratory study. *Cognition and Instruction, 26*(4), 430–511.

Hill, H. C., Rowan, B. and Ball, D. L. (2005). Effects of teachers' mathematical knowledge for teaching on student achievement. *American Educational Research Journal, 42*(2), 371–406.

Hogendijk, J. P. (1991). Desargues' Brouillon Project and the Conics of Apollonius. *Centaurus, 34*(1), 1–43.

Hölzl, R. (1996). How does "dragging" affect the learning of geometry. *International Journal of Computers for Mathematical Learning, 1*(2), 169–187.

Horgan, J. (1993). The death of proof. *Scientific American 269*, 92–103.

Houdement, C. and Kuzniak, A. (2003). Elementary geometry split into different geometrical paradigms. In M. A. Mariotti (Ed.), *Proceedings of CERME 3*. Bellaria: Italy. Retrieved from http://www.mathematik.uni-dortmund.de/~erme/CERME3/ on November 30, 2016.

Houdement, C. and Kuzniak, A. (2006). Paradigmes géométriques et enseignement de la géométrie. *Annales de Didactique et de Sciences Cognitives, 11*, 175–193.

Howson, A. G. (1982). *A history of mathematics education in England*. Cambridge: Cambridge University Press.

Howson, G. and Rogers, L. (2014). Mathematics education in the United Kingdom. In A. Karp and G. Schubring (Eds), *Handbook on the history of mathematics education*, (pp. 257–282). New York: Springer.

Hoyles, C., Foxman, D. and Küchemann, D. (2002). *A comparative study of geometry curricula*. Qualifications and Curriculum Authority, London.

Hoyles, C. and Jones, K. (1998) Proof in dynamic geometry contexts. In, C. Mammana and V. Villani (Eds) *Perspectives on the teaching of geometry for the 21st century: An ICMI study* (pp. 121–128). Dordrecht, Netherlands: Kluwer.

Hoyles, C. and Noss, R. (2003). What can digital technologies take from and bring to research in mathematics education? In A. J. Bishop, M. A. Clements, C. Keitel, J. Kilpatrick and F. K.S. Leung (Eds), *Second international handbook of mathematics education*. (pp. 323–349). Dordrecht, The Netherlands: Kluwer.

Hsu, H.Y. (2010). The study of Taiwanese students' experiences with geometric calculation with number (GCN) and their performance on GCN and geometric proof (GP). Unpublished doctoral dissertation. University of Michigan, Ann Arbor.

Hsu, H. Y. and Silver, E. A. (2014). Cognitive complexity of mathematics instructional tasks in a Taiwanese classroom: An examination of task sources. *Journal for Research in Mathematics Education, 45*(4), 460–496.

Jackson, G. B. (1924). *The teaching of geometry in secondary schools.* Manchester University, MEd Thesis.

Jahnke, H. N. (2007). Proofs and hypotheses. *Zentralblatt für Didaktik der Mathematik, 39*(1–2), 79–86.

Jahnke, H. N. and Wambach, R. (2013). Understanding what a proof is: A classroom-based approach. *ZDM, The International Journal on Mathematics Education 45*(3), 469–482.

Johnson, P. (2002). *The renaissance: A short history*. New York: Random House.

Johnson-Laird, P. N. (1983). *Mental models: Towards a cognitive science of language, inference, and consciousness.* Cambridge: Cambridge University Press.

Johnston-Wilder, S. and Mason, J. (Eds). (2005). *Developing thinking in geometry*. London: Chapman.

Jones, K. (2000). Providing a foundation for deductive reasoning: students' interpretations when using dynamic geometry software and their evolving mathematical explanations. *Educational Studies in Mathematics, 44*(1–2), 55–85.

Jones, K. (2002). Issues in the teaching and learning of geometry. In L. Haggarty (Ed.), *Aspects of Teaching Secondary Mathematics: perspectives on practice,* pp. 121–139). London: RoutledgeFalmer.

Jones, K., Fujita, T. and Kunimune, S. (2012). Representations and reasoning in 3-D geometry in lower secondary school. In T. Y. Tso (Ed.), *Proceedings of the 36th Conference of the International Group for the Psychology of Mathematics Education* (Vol. 2, pp. 339–346). Taipei, Taiwan: PME.

Jones, K., Mackrell, K. and Stevenson, I. (2010). Designing digital technologies and learning activities for different geometries. In C. Hoyles and J.-B. Lagrange (Eds), *Mathematics education and technology: Rethinking the terrain* (pp. 47–60). New York: Springer.

Jones, K. and Mooney, C. (2003), Making space for geometry in primary mathematics. In I. Thompson (Ed.). *Enhancing primary mathematics teaching,* (pp. 3–15). London: Open University Press.

Jones, K., Mooney, C. and Harries, T. (2002). Trainee primary teachers' knowledge of geometry for teaching. *Proceedings of the British Society for Research into Learning Mathematics, 22*(2), 95–100.

Kapadia, R. (1980) Euclid must go. *International Journal of Mathematical Education in Science and Technology, 11*(3), 327–329.

Kapur, J. N. (1990). *Fascinating world of Indian mathematics,* Vol. 7. New Delhi: Mathematical Sciences Trust.

Kaput, J. J. (1986). Information technology and mathematics: opening new representational windows. *The Journal of Mathematical Behavior, 5,* 187–207.

Karabenick, S. A., Woolley, M. E., Friedel, J. M., Ammon, B. V., Blazevski, J., Bonney, C. R., . . . and Kelly, K. L. (2007). Cognitive processing of self-report items in educational research: Do they think what we mean? *Educational Psychologist, 42*(3), 139–151.

Karmiloff-Smith, A. (1990). Constraints on representational change: Evidence from children's drawing. *Cognition, 34*(1), 57–83.

Karp, A. and Werner, A. (2011). On the teaching of geometry in Russia. In A. Karp and B. Vogeli (Eds) *Russian mathematics education: Programs and practices* (pp.81–128). London: World Scientific

Kaufmann, H., Schmalstieg, D. and Wagner, M. (2000). Construct3D: A virtual reality application for mathematics and geometry education. *Education and Information Technologies, 5*(4), 263–276.

Kaur, H. (2015). Two aspects of young children's thinking about different types of dynamic triangles: prototypicality and inclusion. *ZDM The International Journal on Mathematics Education, 47*(3), 407–420.

Kennedy, H.C. (1972). The origins of modern axiomatics: Pasch to Peano. *The American Mathematical Monthly, 79*(2), 133–136.

Kenney, M. J. and Hirsch, C. R. (1991). Discrete Mathematics across the Curriculum, K-12. *1991 Yearbook*. Reston, VA: National Council of Teachers of Mathematics.

Klein, F. (1872). *Vergleichende Betrachtungen über neuere geometrische Forschungen*. Erlangen: Deichert.

Klein, F. (1893). A comparative review of recent researches in geometry. *Bulletin of the American Mathematical Society, 2*(10), 215–249.

Klein, F. (1895). Über die Beziehungen der neueren Mathematik zu den Anwendungen. *Zeitschrift für mathematischen und naturwissenschaftlichen Unterricht* 26, 535–540.

Klein, F. (1907). *Vorträge über den mathematischen Unterricht an den höheren Schulen (Vol. 1)*. BG Leipzig: Teubner.

Klein, F. (1925). *Elementarmathematik vom höheren Standpunkte aus (Vol. 2: Geometrie*. Berlin: Springer.

Klein, F. (1939). *Elementary mathematics from an advanced standpoint: Geometry*. New York: Dover.

Kline, M. (1983). *Mathematics: The loss of certainty*. Oxford, Great Britain: Oxford University Press.

Knight, K. C. (2006). An investigation into the change in the Van Hiele levels of understanding geometry of pre-service elementary and secondary mathematics teachers. Unpublished doctoral dissertation: University of Maine.

Knoche, N., Lind, D., Blum, W., Cohors-Fresenborg, E., Flade, L., Löding, W. . . . and Wynands, A. (2002). (Deutsche PISA-Expertengruppe Mathematik, PISA-2000) Die PISA-2000-Studies, einige Ergebnisse und Analysen. *Journal für Mathematik-Didaktik, 23*(3–4), 159–202.

Knuth, E. J. (2002a). Teachers' conceptions of proof in the context of secondary school mathematics. *Journal of Mathematics Teacher Education, 5*(1), 61–88.

Knuth, E. J. (2002b). Secondary school mathematics teachers' conceptions of proof. *Journal for Research in Mathematics Education, 33*(5), 379–405.

Koedinger, K. R. and Anderson, J. R. (1990). Abstract planning and perceptual chunks: Elements of expertise in geometry. *Cognitive Science, 14*, 511–550.

Kondo, Y., Fujita, T., Jones, K., Kunimune, S. and Kumakura, H. (2014, July). Students' level of 3D geometrical thinking. In P. Liljedahl, C. Nicol, S. Oesterle, & D. Allen (Eds.), *Proceedings of the 38th Conference of the International Group for the Psychology of Mathematics Education*. (vol. 3 pp. 25–32) Vancouver: Canada: Simon Fraser University.

Koschmann, T. and Mori, J. (2016). "It's understandable enough, right?" The natural accountability of a mathematics lesson. *Mind, Culture, and Activity, 23* (1), 65–91.

Koseki, K. (Ed.). (1987). *The teaching of geometrical proof.* Tokyo: Meiji Tosho Publishers (in Japanese).

Kospentaris, G., Spyrou, P. and Lappas, D. (2011). Exploring students' strategies in area conservation geometrical tasks. *Educational Studies in Mathematics*, 77(1), 105–127.

Kotsopoulos, D., Cordy, M. and Langemeyer, M. (2015). Children's understanding of large-scale mapping tasks: an analysis of talk, drawings, and gesture. *ZDM The International Journal on Mathematics Education*, 47(3), 451–463.

Krüger, K. (2000). Erziehung zum funktionalen Denken Zur Begriffsgeschichte eines didaktischen Prinzips. *Journal für Mathematik-Didaktik, 21*(3–4), 326–327.

Küchemann, D. and Hoyles, C. (2002). The quality of students' reasons for the steps in a geometric calculation. *Proceedings of the British Society for Research into Learning Mathematics, 22*(2), 43–48.

Kuzniak, A. and Rauscher, J. C. (2011). How do teachers' approaches to geometric work relate to geometry students' learning difficulties? *Educational Studies in Mathematics*, 77(1), 129–147.

Laborde, C. (1998). Geometry behind the French national curricula in the last decades. In Mammana, C. and Villani, V. (Eds.) *Perspectives on the teaching of geometry for the 21st century: An ICMI study* (pp. 214–222). Dordrecht, The Netherlands: Kluwer.

Laborde, C. (1998). Visual phenomena in the teaching/learning of geometry in a computer-based environment. In C. Mammana and V. Villani (Eds) *Perspectives on the teaching of geometry for the 21st century: An ICMI study* (pp. 113–121). Dordrecht, The Netherlands: Kluwer.

Laborde, C. (2000, October). Géométrie–période 2000 et apres. One hundred years of l'enseignement mathématique, moments of mathematics education in the twentieth century. *Proceedings of the EM–ICMI Symposium*. Geneva. *Geneva.*

Laborde, C. (2002). Integration of technology in the design of geometry tasks with Cabri-Geometry. *International Journal of Computers for Mathematical Learning, 6*(3), 283–317.

Laborde, C. (2005). The hidden role of diagrams in students' construction of meaning in geometry. In J. Kilpatrick, C. Hoyles, and O. Skovsmose, with P. Valero (Eds), *Meaning in mathematics education* (pp. 159–179). New York: Springer.

Laborde, C. (2007). The role and uses of technologies in mathematics classrooms: Between challenge and modus vivendi. *Canadian Journal of Math, Science & Technology Education*, 7(1), 68–92.

Laborde, C. (2008). Experiencing the multiple dimensions of mathematics with dynamic 3D geometry environments: Illustration with Cabri 3D. *The Electronic Journal of Mathematics and Technology, 2*(1), 38–53.

Laborde, C. (2008). Multiple dimensions involved in the design of tasks taking full advantage of dynamic interactive geometry. *Tecnologias e Educação Matemática*, 36–50.

Laborde, C., Kynigos, C., Hollebrands, K. and Strässer, R. (2006). Teaching and learning geometry with technology. In Gutiérrez, A. and Boero, P. (Eds.), *Handbook of research on the psychology of mathematics education: Past, present and future* (pp. 275–304). Rotterdam: Sense Publishers.

Lacroix, S. F. (1797). *Traité du calcul différentiel et du calcul integral* (Vols 1–3). Paris: Duprat, J.B.M.

Lagrange, J. B., Artigue, M., Laborde, C. and Trouche, L. (2003). Technology and mathematics education: A multidimensional study of the evolution of research and innovation. In A. Bishop, M. A. Clements, C. Keitel, J. Kilpatrick and F. Leung (Eds) *Second international handbook of mathematics education* (pp. 237–269). Dordrecht: Kluwer.

Lakatos, I. (1976). *Proofs and refutations: The logic of mathematical discovery*. Cambridge University Press.

Lampert, M. (2001). *Teaching problems and the problems of teaching*. New Haven, CT: Yale University Press.

Landau, B. (1988). The construction and use of spatial knowledge in blind and sighted children. In J. Stiles, M. Kritchevsky, and U. Bellugi (Eds). *Spatial cognition: Brain bases and development* (pp. 343–371). Hillsdale, NJ: Erlbaum.

Larkin, J. and Simon, H. (1987). Why a diagram is (sometimes) worth ten thousand words. *Cognitive Science, 11*(1), 65–100.

Lee, S. A., Sovrano, V. A. and Spelke, E. S. (2012). Navigation as a source of geometric knowledge: Young children's use of length, angle, distance, and direction in a reorientation task. *Cognition, 123*(1), 144–161.

Legendre, A.-M. (1843). *Elements of geometry and trigonometry* (translated by D. Brewster; revised by C. Davies). Philadelphia: AS Barnes.

Lehrer, R., Jacobson, C., Thoyre, G., Kemeny, V., Strom, D., Horvath, J., Gance, S. and Koehler, M. (1998). Developing understanding of geometry and space in the primary grades. In R. Lehrer and D. Chazan (Eds), *Designing learning environments for developing understanding of geometry and space,* (pp. 169–200). Mahwah, NJ: Erlbaum.

Lehrer, R., Jenkins, M. and Osana, H. (1998). Longitudinal study of children's reasoning about space and geometry. In R. Lehrer and D. Chazan (Eds), *Designing learning environments for developing understanding of geometry and space,* (pp. 137–167). Mahwah, NJ: Erlbaum.

Lemke, J. (1990). *Talking science: Language, learning and values.* Norwood, NJ: Ablex.

Leung, A. (2003). Dynamic geometry and the theory of variation. In N. A. Pateman, B. J. Doughherty, and J. T. Zillox (Eds.), *Proceedings of PME 27: Psychology of Mathematics Education 27th International Conference, Volume 3* (pp. 197–204). Honolulu: University of Hawaii.

Leung, A. (2008). Dragging in a dynamic geometry environment through the lens of variation. *International Journal of Computers for Mathematical Learning, 13*(2), 135–157.

Leung, A., Chan, Y.C. and Lopez-Real, F. (2000). Instrumental genesis in dynamic geometry environments. *Learning, 44,* 1–161.

Levenson, E., Tirosh, D. and Tsamir, P. (2011). *Preschool geometry: Theory, research, and practical perspectives.* Rotterdam: Sense.

Lewis, C. C. (2002). *Lesson study: A handbook of teacher-led instructional change.* Philadelphia, PA: Research for Better Schools.

Liben, L. S. (2007). Education for spatial thinking. *Handbook of child psychology, Vol 4: Child psychology research in practice* (K. A. Renninger and I. Siegel, Eds, pp. 197–247). Hoboken, NJ: Wiley.

Liben, L. S., Kastens, K. A. and Stevenson, L. M. (2002). Real-world knowledge through real-world maps: A developmental guide for navigating the educational terrain. *Developmental Review, 22*(2), 267–322.

Linn, M. C. and Petersen, A. C. (1985). Emergence and characterization of sex differences in spatial ability: A meta-analysis. *Child Development, 56*(6), 1479–1498.

Linn, M. C. and Petersen, A. C. (1986). A meta-analysis of gender differences in spatial ability: Implications for mathematics and science achievement. In J. S. Hyde and M.C. Linn (Eds), *The psychology of gender: Advances through meta-analysis* (pp. 7–101). Baltimore, MD: The Johns Hopkins University Press.

Lortie, D. C. (1975). *Schoolteacher: A sociological inquiry.* Chicago: University of Chicago Press.

Mammana, C. and Villani, V. (1994). Perspectives on the teaching of geometry for the 21st century (discussion document for an ICMI study). *L'Enseignement Mathématique, 40*(3–4), 345–357.

Manders, K. (2008). The Euclidean diagram. In P. Mancosu (Ed.), *The Philosophy of Mathematical Practice* (pp. 80–133). Oxford: OUP. (Original work dated 1995).

Manders, K. (2008). Diagram based geometric practice. In P. Mancosu (Ed), *The philosophy of mathematical practice* (pp. 65–79). Oxford: OUP.

Mansfield, H. M. and Scott, J. (1990). Young children solving spatial problems. In G. Booker, P. Cobb, and T. N. de Mendicuti (Eds.), Proceedings of the 14th annual conference of the International Group for the Psychology of Mathematics Education (Vol. 2, pp. 275–282). Oaxlepec, Mexico: International Group for the Psychology of Mathematics Education.

Margolinas, C. (1993). *De l'importance du vrai et du faux dans la classe de mathématiques.* Grenoble: La Pensée Sauvage.

Mariotti, M. A. (1995). Images and concepts in geometrical reasoning. In R. Sutherland and J. Mason (Eds), *Exploiting mental imagery with computers in mathematics education* (pp. 97–116). Berlin: Springer.

Mariotti, M. A. (2000). Introduction to proof: The mediation of a dynamic software environment. *Educational Studies in Mathematics, 44*(1–2), 25–53

Mariotti, M. A., Bartolini Bussi, M., Boero P., Ferri, F. and Garuti, R. (1997). Approaching geometry theorems in contexts: from history and epistemology to cognition. In E. Pehkonen

(Ed.), Proceedings of the 21st conference of the International Group for the Psychology of Mathematics Education, Vol. 1 (pp. 180–195). Lahti, Finland: University of Helsinki.

Mayberry, J. (1983). The van Hiele levels of geometric thought in undergraduate preservice teachers. *Journal for Research in Mathematics Education, 14*(1), 58–69.

McGowen, M. and Tall, D. (2010). Metaphor or met-before? The effects of previous experience on the practice and theory of learning mathematics. *Journal of Mathematical Behavior, 29*, 169–179.

McKenney, S. and Reeves, T. C. (2012). *Conducting educational design research*. New York: Routledge.

McKnight, C. C. (1987). *The underachieving curriculum: Assessing US school mathematics from an international perspective. A national report on the Second International Mathematics Study*. Champaign, IL: Chester: Stipes Publishing.

Mehan, H. (1979). *Learning lessons: Social organization in the classroom*. Cambridge, MA: Harvard University Press.

Mehan, H. and Wood, H. (1975). *The reality of ethnomethodology*. New York: Wiley.

Menghini, M. (2006). The role of projective geometry in Italian education and institutions at the end of the 19th century. *International Journal for the History of Mathematics Education, 1*(1), 35–55.

Meserve, B. E. (1959). The education of mathematics teachers. *American Mathematical Monthly, 66*(10), 909–911.

Mesquita, A. L. (1998). On conceptual obstacles linked with external representations in geometry. The *Journal of Mathematical Behavior, 17*(2), 183–196.

Miller, G. A. (2003). The cognitive revolution: A historical perspective. *Trends in Cognitive Sciences*, 7(3), 141–144.

Mingus, T. T. and Grassl, R. M. (1999). Preservice teacher beliefs about proofs. *School Science and Mathematics, 99*(8), 438–444.

Miyazaki, M., Fujita, T. and Jones, K. (2015). Flow-chart proofs with open problems as scaffolds for learning about geometrical proofs. *ZDM The International Journal on Mathematics Education*, 47(7), 1211–1224.

Miyazaki, M., Kimiho, C., Katoh, R., Arai, H., Ogihara, F., Oguchi, Y. . . . and Komatsu, K. (2012). Potentials for spatial geometry curriculum development with three-dimensional dynamic geometry software in lower secondary mathematics. *International Journal for Technology in Mathematics Education, 19*(2), 73–79.

Moise, E. (1974). *Elementary geometry from an advanced standpoint*. Reading, MA: Addison Wesley.

Moise, E. and Downs, F. (1964). *Geometry*. Reading, MA: Addison Wesley.

Morgan, C. and Alshwaikh, J. (2012). Communicating experience of 3D space: Mathematical and everyday discourse. *Mathematical Thinking and Learning, 14*(3), 199–225.

Moustakas, K., Strintzis, M. G., Tzovaras, D., Carbini, S., Bernier, O., Viallet, J. E., Raidt, R., Mancas, M., Dimiccoli, M., Yagci, E., Balci, S. and Leon, E. I. (2006). Masterpiece: physical interaction and 3D content-based search in VR applications. *MultiMedia, IEEE, 13*(3), 92–100.

Moyer, P. S. (2001). Are we having fun yet? How teachers use manipulatives to teach mathematics. *Educational Studies in Mathematics*, 47(2), 175–197.

Mueller, I. (1970). Aristotle on geometrical objects. *Archiv für Geschichte der Philosophie, 52*(2), 156–171.

Mueller, I. (1981). *Philosophy of mathematics and deductive structure in Euclid's Elements*. Cambridge, MA: The MIT Press.

Mueller, I. (1991). On the notion of a mathematical starting-point in Plato, Aristotle, and Euclid. *Bowen, 8*(4), 59–97.

Mulhern, F. and Rae, G. (1998). Development of a shortened form of the Fennema-Sherman Mathematics Attitudes Scales. *Educational and Psychological Measurement, 58*(2), 295–306.

Müller, T. (2006). *Die Bedeutung neuer Medien in der Fachdidaktik für den Unterrichtsgegenstand Darstellende Geometrie.* Technischen Universität Wien. Austria.

Mullis, I. V., Martin, M. O., Gonzalez, E. J. and Chrostowski, S. J. (2004). *TIMSS 2003 International Mathematics Report: Findings from IEA's Trends in International Mathematics and Science Study at the Fourth and Eighth Grades.* Chestnut Hill, MA: TIMSS & PIRLS International Study Center.

Nachlieli, T. and Herbst, P. with González, G. (2009). Seeing a colleague encourage a student to make an assumption while proving: What teachers put to play in casting an episode of geometry instruction. *Journal for Research in Mathematics Education, 40*(4), 427–459.

Nagel, E. (1939). The formation of modern conceptions of formal logic in the development of geometry. *Osiris*, 7, 142–223.

National Council of Teachers of Mathematics (1989). *Curriculum and evaluation standards for school mathematics.* Reston, VA: Author.

National Council of Teachers of Mathematics. Commission on Teaching Standards for School Mathematics. (1991). *Professional standards for teaching mathematics.* Reston, VA: Author.

National Council of Teachers of Mathematics (2000). *Principles and standards for school mathematics.* Reston, VA: Author.

National Education Association of the United States. Committee of Ten on Secondary School Studies. (1894). *Report of the Committee of Ten on Secondary School Studies: With the reports of the conferences arranged by the committee.* New York: American Book Company and National Education Association.

Netz, R. (1998a). Greek mathematical diagrams: Their use and their meaning. *For the Learning of Mathematics, 18*(3), 33–39.

Netz, R. (1998b). Deuteronomic texts: late antiquity and the history of mathematics. *Revue d'Histoire des Mathématiques, 4*(2), 261–288.

Netz, R. (2002). Proof, amazement, and the unexpected. *Science, 298*, 967–968.

Newcombe, N. S. and Huttenlocher, J. (2000). *Making space. The development of spatial representation and reasoning.* Cambridge, MA: MIT Press.

Ng, D. (2011). Indonesian primary teachers' mathematical knowledge for teaching geometry: implications for educational policy and teacher preparation programs. *Asia-Pacific Journal of Teacher Education, 39*(2), 151–164.

Ng, O. L. and Sinclair, N. (2015). Young children reasoning about symmetry in a dynamic geometry environment. *ZDM The International Journal on Mathematics Education, 47*(3), 421–434.

Nkwe, D. T. (1985). Teachers attitude, with special references to mathematics teaching. Unpublished MEd Dissertation, University of Witwatersrand, Johannesburg, South Africa.

O'Connor, J. J. and Robertson, E. F. (2004). *Sylvestre François Lacroix.* Retrieved on January 15, 2014 from http://www-history.mcs.st-and.ac.uk/Printonly/Lacroix.html

OECD - Organisation for Economic Co-operation and Development (Ed.) (1999). *Measuring Student Knowledge and Skills: The PISA 2000 Assessment of Reading, Mathematical and Scientific Literacy. Education and Skills.*

OECD - Organisation for Economic Co-operation and Development (Ed.) (2003). *The PISA assessment framework: Mathematics, reading, science and problem solving knowledge and skills.*

OECD - Organisation for Economic Cooperation and Development (2014). *PISA 2012 technical report.* Paris: OECD.

Oldknow, A. and Tetlow, L. (2008). Using dynamic geometry software to encourage 3D visualization and modelling. *The Electronic Journal of Mathematics and Technology, 2*(1), 54–61.

Osta, I., Laborde, C., Hoyles, C., Jones, K., Graf, K.D. and Hodgson, B. (1998). Computer technology and the teaching of geometry. In C. Mammana and V. Villani (Eds), *Perspectives on the Teaching of Geometry for the 21st Century: An ICMI study.* (pp. 109–158). Dordrecht, The Netherlands: Kluwer.

O'Toole, A. L. (1941), College geometry for secondary school teachers. *American Mathematical Monthly*, *48*(5), 319–326.

Özerem, A. (2012). Misconceptions in geometry and suggested solutions for seventh grade students. *Procedia-Social and Behavioral Sciences*, *55*, 720–729.

Pajares, M. F. (1992). Teachers' beliefs and educational research: Cleaning up a messy construct. *Review of Educational Research*, *62*(3), 307–332.

Pajares, F. and Miller, M. D. (1994). Role of self-efficacy and self-concept beliefs in mathematical problem solving: A path analysis. *Journal of Educational Psychology*, *86*(2), 193.

Panza, M. (2012). The twofold role of diagrams in Euclid's plane geometry. *Synthese*, *186*(1), 55–102.

Parzysz, B. (1988). 'Knowing' vs. 'seeing': Problems of the plane representation of space geometry figures. *Educational Studies in Mathematics*, *19*(1), 79–92.

Pasch, M. (1882). Vor*lesungen über neuere Geometrie*. Leipzig: Teubner.

Pea, R. D. (1985). Beyond amplification: Using the computer to reorganize mental functioning. *Educational Psychologist*, *20*(4), 167–182.

Pegg, J. and Tall, D. (2005). The fundamental cycle of concept construction underlying various theoretical frameworks. *Zentralblatt für Didaktik der Mathematik*, *37*(6), 468–475.

Peirce, C. S. (1955). Logic as semiotic: The theory of signs. In J. Buchler (Ed.), *Philosophical writings of Peirce* (pp. 98–119). New York: Dover. (Original work 1897).

Peitgen, H. O., Jürgens, H., Saupe, D. and Maletsky, E. (1992). *Fractals for the classroom* (Vol. 1). New York: Springer-Verlag.

Perry, J. (1902), The teaching of mathematics, *Educational Review*, *23*, 158–189.

Piaget, J. and Inhelder, B. (1956). *The child's conception of space*. London: Routledge and Kegan Paul.

Piaget, J. and Inhelder, B. (1963). *The child's conception of space* (Translated from the French by F. J. Lagdon and J. L. Lunzer) Oxon, England: Routledge & Kegan Paul.

Piaget, J., lnhelder, B. and Szeminska, A. (1960). *The child's conception of geometry*. London: Routledge and Kegan Paul.

Pittalis, M. and Christou, C. (2013). Coding and decoding representations of 3D shapes. *The Journal of Mathematical Behavior*, *32*(3), 673–689.

Plake, B.S. and Parker, C.S. (1982). The development and validation of a revised version of the mathematics anxiety rating scale. *Educational and Psychological Measurement*, *42*(2), 551–557.

Playfair, J. (1848). *Elements of geometry: Containing the first six books of Euclid, with a supplement on the quadrature of the circle, and the geometry of solids to which are added, elements of plane and spherical trigonometry*. New York: WE Dean.

Poincaré, H. (1904). Les définitions générales en mathématiques [General definitions in mathematics]. *L'Enseignement Mathématique 5*, 257–283.

Pölya, G. (1957) *How to solve it: A new aspect of mathematical method*, 2nd edition. Princeton: Princeton University Press.

Proclus (1970). *A commentary on the first book of Euclid's* Elements. Translated with introduction and notes by Glenn R. Morrow. Princeton: Princeton University Press.

Radford, L. (2014). The progressive development of early embodied algebraic thinking. *Mathematics Education Research Journal*, *26*(2), 257–277.

Rauscher, J.-C. (1993). *L'hétérogénéité des professeurs face à des élèves hétérogènes. Le cas de l'enseignement de la géométrie au début du collège*. Strasbourg: University of Strasbourg.

Reid, D. A. and Knipping, C. (2010). *Proof in mathematics education: Research, learning and teaching*. Rotterdam, The Netherlands: Sense Publishers.

Richards, J. L. (1988). *Mathematical visions: The pursuit of geometry in Victorian England*. London: Academic Press.

Richardson, F. C. and Suinn, R. M. (1972). The Mathematics Anxiety Rating Scale: Psychomet *Counseling Psychology*, *19*(6), 551.

Roberts, D. L. (2014). History of tools and technologies in mathematics education. In A. Karp and G. Schubring (Eds), *Handbook on the history of mathematics education.* (pp. 565–578). New York: Springer.

Roberts, S. (2006). *King of infinite space: Donald Coxeter, the man who saved geometry.* New York: Walker & Company.

Rosch, E. H. (1973). On the internal structure of perceptual and semantic categories, In T. E. Moore (Ed.), *Cognitive development and the acquisition of language,* (pp. 111–144). Oxford, England: Academic Press.

Rosenberg, H. (1958). The changing concept of trigonometry as a school subject. *The Mathematics Teacher, 51*(4) 246–252.

Roth, W. M. and Maheux, J. F. (2015). The stakes of movement: A dynamic approach to mathematical thinking. *Curriculum Inquiry, 45*(3), 266–284.

Rowland, T., Huckstep, P. and Thwaites, A. (2005). Elementary teachers' mathematics subject knowledge: The knowledge quartet and the case of Naomi. *Journal of Mathematics Teacher Education, 8*(3), 255–281.

Rowland, T. and Ruthven, K. (Eds). (2011). *Mathematical knowledge in teaching.* New York: Springer.

Royce, J. (1905). Kant's Doctrine of the Basis of Mathematics. *The Journal of Philosophy, Psychology and Scientific Methods, 2*(8), 197–207.

Runesson, U. (2006). What is it possible to learn? On variation as a necessary condition for learning. *Scandinavian Journal of Educational Research, 50*(4), 397–410.

Ruthven, K., Hennessy, S. and Deaney, R. (2008). Constructions of dynamic geometry: A study of the interpretative flexibility of educational software in classroom practice. *Computers & Education, 51*(1), 297–317.

Saccheri, G. (2014). *Euclid vindicated from every blemish: Edited and annotated by Vincenzo De Risi.* Translated by G. B. Halsted and L. Allegri. Cham, Switzerland: Springer.

Sack, J. J. (2013). Development of a top-view numeric coding teaching-learning trajectory within an elementary grades 3-D visualization design research project. *The Journal of Mathematical Behavior, 32*(2), 183–196.

Sarama, J. and Clements, D. H. (2002). Building blocks for young children's mathematical development. *Journal of Educational Computing Research, 27*(1), 93–110.

Sarama, J. and Clements, D. (2009). *Early childhood mathematics education research: Learning trajectories for young children.* New York: Routledge.

Saye, J. W. (1998). Technology in the classroom: The role of disposition in teacher gate keeping. *Journal of Curriculum and Supervision, 13*(3), 210–234.

Scher, D. (2000). Lifting the curtain: The evolution of the Geometer's Sketchpad. *The Mathematics Educator, 10*(1), 42–48.

Schmidt, W. H., Blömeke, S., Tatto, M. T., Hsieh, F. J., Cogan, L., Houang, R. T. and Schwille, J. (2011). *Teacher education matters: A study of middle school mathematics teacher preparation in six countries.* New York, NY: Teachers College Press.

Schmidt, W. H., Houang, R. T., Cogan, L., Blömeke, S., Tatto, M. T., Hsieh, F. J. . . . and Paine, L. (2008). Opportunity to learn in the preparation of mathematics teachers: its structure and how it varies across six countries. *ZDM The International Journal on Mathematics Education, 40*(5), 735–747.

Schoenfeld, A. H. (1988). When good teaching leads to bad results: The disasters of' well-taught mathematics courses. *Educational Psychologist, 23*(2), 145–166.

Schoenfeld, A. H. (1989). Explorations of students' mathematical beliefs and behavior. *Journal for Research in Mathematics Education, 20*(4), 338–355.

Schoenfeld, A. H. (1994). *Mathematical thinking and problem solving.* Hillsdale, NJ: Erlbaum.
Schopenhauer (1864). *Über die vierfache Wurzel des Satzes vom zureichenden Grunde.* Leipzig: FA Brockhaus.
Schreiber, P. and Brentjes, S. (1987). *Euklid.* Leipzig: Teubner.
Schubring, G. (2014). Mathematics education in germany (Modern Times). In A. Karp and G. Schubring (Eds), *Handbook of the history of mathematics education* (pp. 241–255). New York: Springer.
Schweiger, F. (2006). Fundamental ideas—a bridge between mathematics and mathematical education. In J. Maaß, and W. Schlöglmann (Eds), *New mathematics education research and practice* (pp. 63–74). Rotterdam: Sense Publishers.
Scriba, C. J. and Schreiber, P. (2015). *5000 years of geometry: mathematics in history and culture.* Basel: Springer.
Senk, S. L. (1989). Van Hiele levels and achievement in writing geometry proofs. *Journal for Research in Mathematics Education, 10*(3), 309–321.
Sfard, A. (2008). *Thinking as communicating: Human development, the growth of discourses, and mathematizing.* Cambridge, England: Cambridge University Press.
Shafer, K. G. (2004). Two high school teachers' initial use of Geometer's Sketchpad: Issues of implementation. Unpublished doctoral dissertation, Western Michigan University.
Shulman, L. S. (1986a). Paradigms and research programs in the study of teaching: A contemporary perspective. In M. Wittrock (Ed.), *Handbook of research on teaching,* 3rd edition (pp. 3–36). New York: Macmillan.
Shulman, L. S. (1986b). Those who understand: Knowledge growth in teaching. *Educational Researcher, 15*(2), 4–14.
Silverman, J. and Thompson, P. W. (2008). Toward a framework for the development of mathematical knowledge for teaching. *Journal of Mathematics Teacher Education, 11*(6), 499–511.
Sinclair, N. (2008). *History of the geometry curriculum in the United States.* Charlotte, NC: Information Age.
Sinclair, N. and Bruce, C. D. (2015). New opportunities in geometry education at the primary school. *ZDM The International Journal on Mathematics Education, 47*(3), 319–329.
Sinclair, N., Cirillo, M. and de Villiers, M. (in press). The teaching and learning of geometry. In J. Cai (Ed.), *First Compendium for research in mathematics education.* Reston, VA: National Council of Teachers of Mathematics.
Sinclair, N. and Moss, J. (2012). The more it changes, the more it becomes the same: The development of the routine of shape identification in dynamic geometry environment. *International Journal of Educational Research, 51,* 28–44.
Sinclair, N. and Robutti, O. (2013). Technology and the role of proof: The case of dynamic geometry. In M. A. Clements, A. Bishop, C. Keitel, J. Kilpatrick and F. Leung (Eds), *Third international handbook of mathematics education* (pp. 571–596). New York: Springer.
Sinclair, N. and Yurita, V. (2008). To be or to become: How dynamic geometry changes discourse. *Research in Mathematics Education, 10*(2), 135–150.
Smith, J. T. (2011). *Methods of geometry.* New York: John Wiley & Sons.
Smith, N. D. (1999). Images, education, and paradox in Plato's Republic. *Apeiron, 32*(4), 125–142.
Sneed, J. D. (1971). *The logical structure of mathematical physics.* Dordrecht, The Netherlands: Reidel.
Sörbom, G. (1994). Gombrich on the Greek art revolution. *The Nordic Journal of Aesthetics,* 7(12), 62–77.
Spencer, W. (1876). *Inventional geometry.* Woodstock, GA: American Book Company.

Spikol, D. and Eliasson, J. (2010, April). Lessons from designing geometry learning activities that combine mobile and 3D tools. In *Wireless, Mobile and Ubiquitous Technologies in Education (WMUTE), 2010 6th IEEE International Conference* (pp. 137–141). IEEE.

Stamper, A. W. (1909). *A history of the teaching of elementary geometry: With reference to present-day problems*. New York: Teachers College, Columbia University.

Stanton, M. and Moore-Russo, D. (2012). Conceptualizations of slope: A review of state standards. *School Science and Mathematics, 112*(5), 270–277.

Stavy, R. and Tirosh, D. (2000). *How students (mis-) understand science and mathematics: Intuitive rules*. New York: Teachers College Press.

Steele, M. (2006) Middle grades geometry and measurement: Examining change in knowledge needed for teaching through a practice-based teacher education experience. Unpublished doctoral dissertation, University of Pittsburgh.

Steele, M. and Hillen, A. (2012). The content-focused methods course: A model for integrating pedagogy and mathematics content. *Mathematics Teacher Educator 1*(1), 53–70

Steffe, L. (1994). Children's multiplying schemes. In G. Harel and J. Confrey (Eds). *The development of multiplicative reasoning in the learning of mathematics* (pp. 3–40). Albany, NY: SUNY Press.

Stegmüller, W. (1985). *Probleme und Resultate der Wissenschaftstheorie und analytischen Philosophie Band II: Theorie und Erfahrung: Theorienstrukturen und Theoriendynamik* (2nd edition). Berlin, Germany: Springer Verlag.

Stein, M. K., Grover, B. W., and Henningsen, M. (1996). Building student capacity for mathematical thinking and reasoning. *American Educational Research Journal, 33*(2), 455–488.

Stigler, J. W. and Hiebert, J. (1999). *The teaching gap*. New York, NY: Free Press.

Stillwell, J. (1996). *Sources of hyperbolic geometry*. Providence, RI: American Mathematical Society.

Stipek, D. J., Givvin, K. B., Salmon, J. M. and MacGyvers, V. L. (2001). Teachers' beliefs and practices related to mathematics instruction. *Teaching and Teacher Education, 17*(2), 213–226.

Stodolsky, S. S. (1988). *The subject matters: Classroom activity in math and social studies*. Chicago: University of Chicago Press.

Stols, G. and Kriek, J. (2011). Why don't all maths teachers use dynamic geometry software in their classrooms? *Australasian Journal of Educational Technology, 27*(1), 137–151.

Sträßer, R. (2001). CABRI-Géomètre: Does dynamic geometry software (DGS) change geometry and its teaching and learning? *International Journal of Computers for Mathematical Learning, 6*(3), 319–333.

Streeck, J., Goodwin, C. and LeBaron, C. (2011). Embodied interaction in the material world: An introduction. In C. Goodwin, J. Streeck, and C. LeBaron (Eds), *Embodied interaction: language and body in the material world* (pp. 1–26). Cambridge, England: Cambridge University Press.

Struve, H. (1990). *Grundlagen einer Geometriedidaktik*. Mannheim, Bibliographisches Institut.

Stump, S. L. (2001). High school precalculus students' understanding of slope as measure. *School Science and Mathematics, 101*(2), 81–89.

Stylianides, G. J. and Stylianides, A. J. (2008). Proof in school mathematics: Insights from psychological research into students' ability for deductive reasoning. *Mathematical Thinking and Learning, 10*(2), 103–133.

Swafford, J. O., Jones, G. A. and Thornton, C. A. (1997). Increased knowledge in geometry and instructional practice. *Journal for Research in Mathematics Education, 28*(4), 467–483.

Swicegood, G. (2004). The Morley trisector theorem. *The Mathematics Enthusiast, 1*(2), 38–43.

Szabó, Á. (1974). *"Die Muse der Pythagoreer"*. Zur Frühgeschichte der Geometrie. *Historia Mathematica, 1*(3), 291–316.

Szabó, Á. (1978). *The beginnings of Greek mathematics*. Boston: Reidel.

Tabach, M., Levenson, E., Barkai, R., Tirosh, D., Tsamir, P. and Dreyfus, T. (2010). Secondary school teachers' awareness of numerical examples as proof. *Research in Mathematics Education, 12*(2), 117–131.

Tall, D. and Vinner, S. (1981). Concept image and concept definition in mathematics with particular reference to limits and continuity. *Educational Studies in Mathematics, 12*(2), 151–169.

Tam, H. P. and Chen, Y. L. (2012). A regional survey of Taiwan students' performance in geometric construction, In T. Y. Tso (Ed.). *Proceedings of 36th Conference of the International Group for the Psychology of Mathematics Education* (Vol 4, pp. 131–38). Taipei, Taiwan: PME.

Tatto, M. T. and Senk, S. (2011). The mathematics education of future primary and secondary teachers: Methods and findings from the Teacher Education and Development Study in Mathematics. *Journal of Teacher Education, 62*(2), 121–137.

Taylor, R. (1979). Forward to basics. *School Science and Mathematics, 79*(7), 547–554.

Thom, J. S. and McGarvey, L. M. (2015). The act and artifact of drawing(s): observing geometric thinking with, in, and through children's drawings. *ZDM The International Journal on Mathematics Education, 47*(3), 465–481.

Thom, R. (1973). Modern mathematics: Does it exist. In A. G Howson (Ed.), *Developments in mathematics education. Proceedings of the Second International Congress on Mathematical Education.* (pp. 194–209). New York: Cambridge University Press.

Thompson, A. G. (1992). Teacher's beliefs and conceptions: A synthesis of the research. In D. A. Grouws (Ed.), *Handbook of research on mathematics teaching and learning* (pp. 127–146). New York: Macmillan.

Tirosh, D., Stavy, R. and Tsamir, P. (2001). Using the intuitive rules theory as a basis for educating teachers. In F. L. Lin and T. J. Cooney (Eds), *Making sense of mathematics teacher education* (pp. 73–85). Dordrecht, The Netherlands: Kluwer.

Tirosh, D., Tsamir, P., Levenson, E. and Tabach, M. (2011). From preschool teachers' professional development to children's knowledge: Comparing sets. *Journal of Mathematics Teacher Education, 14*(2), 113–131.

Todhunter, I. (1882). *The elements of Euclid, for the use of schools and colleges*. London: Macmillan.

Toepell, M. (2003). Rückbezüge des Mathematikunterrichts und der Mathematikdidaktik in der BRD auf historische Vorausentwicklungen. *Zentralblatt für Didaktik der Mathematik, 35*(4), 177–181.

Torretti, R. (1978). *Philosophy of geometry from Riemann to Poincaré*. Dordrecht, The Netherlands: Reidel.

Trafton, P. R. and LeBlanc, J. F. (1971). Informal geometry in grades K-6. In K. B. Henderson (Ed.) *Geometry in the mathematics curriculum*. (pp. 11–51). Reston, VA: NCTM.

Treffers, A. (1978). *Wiskobas doelgericht* [Wiskobas goal-directed]. Utrecht: IOWO.

Treutlein, P. (1911). *Der geometrische Anschauungsunterricht*. Leipzig / Berlin: Teubner.

Tsamir, P. (2007). When intuition beats logic: prospective teachers' awareness of their same sides–same angles solutions. *Educational Studies in Mathematics, 65*(3), 255–279.

Tsamir, P., Tirosh, D. and Levenson, E. (2008). Intuitive nonexamples: The case of triangles. *Educational Studies in Mathematics, 69*(2), 81–95.

Ulfig, F. (2013). *Geometrische Denkweisen beim Lösen von PISA-Aufgaben*. Wiesbaden: Springer Spektrum.

Ünlü, M., Avcu, S. and Avcu, R. (2010). The relationship between geometry attitudes and self-efficacy beliefs towards geometry. *Procedia-Social and Behavioral Sciences, 9*, 1325–1329.

Usiskin, Z. (1974). Transformation in high school geometry before 1970. *The Mathematics Teacher, 67*(4), 353–360.

Usiskin, Z. (1982). *Van Hiele levels and achievement in secondary school geometry*. Chicago, IL: University of Chicago.

Uttal, D. H., Fisher, J. A. and Taylor, H. A. (2006). Words and maps: developmental changes in mental models of spatial information acquired from descriptions and depictions. *Developmental Science, 9*(2), 221–235.

Van den Akker, J., Gravemeijer, K., McKenney, S. and Nieveen, N. (Eds). (2006). *Educational design research*. New York: Routledge.

Van den Heuvel-Panhuizen, M. (2000). Mathematics education in The Netherlands: A guided tour. *Freudenthal Institute CD-ROM for ICME 9*. Utrecht: Utrecht University.

Van der Sandt, S. and Nieuwoudt, H. D. (2003). Grade 7 teachers' and prospective teachers' content knowledge of geometry. *South African Journal of Education, 23*(3), 199–205.

Van der Sandt, S. and Nieuwoudt, H. D. (2005). Geometry content knowledge: Is pre-service training making a difference? *African Journal of Research in Mathematics, Science and Technology Education, 9*(2), 109–120.

Van der Waerden, B. L. (1967). Klassische und moderne Axiomatik. *Elemente der Mathematik, 22*, 1–4.

Van Hiele, P. M. (1999). Developing geometric thinking through activities that begin with play. *Teaching Children Mathematics, 5*(6), 310–316.

Van Hiele-Geldof, D. (1957/1984). The didactics of geometry in the lowest class of secondary school. In D. Fuys, D. Geddes and R. Tischler, *English translation of selected writings of Dina van Hiele-Geldof and P. M. van Hiele* (pp. 1–214). Brooklyn: Brooklyn College. (Original document in Dutch. De didaktiek van de meetkundein de eerste klas van het V. H. M. O., Unpublished doctoral dissertation, University of Utrecht, 1957).

Verdine, B.N., Golinkoff, R.M., Hirsh-Pasek, K. and Newcombe, N. S. (2014). Finding the missing piece: Blocks, puzzles, and shapes fuel school readiness. *Trends in Neuroscience and Education, 3*(1), 7–13.

Vergnaud, G. (1996). The theory of conceptual fields. In Steffe, L. P., Nesher, P., Cobb, P., Goldin, G. and Greer, B. (Eds) *Theories of mathematical learning*, (pp. 219–239). Hillsdale, NJ: Erlbaum.

Vinner, S. (1991). The role of definitions in the teaching and learning of mathematics. In D. Tall (Ed.), *Advanced mathematical thinking* (pp. 65–81). Dordrecht, The Netherlands: Kluwer.

Vinner, S. and Hershkowitz, R. (1980). Concept images and some common cognitive paths in the development of some simple geometric concepts. In R. Karplus (Ed.), *Proceedings of the 4th international conference for the psychology of mathematics education* (pp. 177–184). Berkeley, CA: Lawrence Hall of Science, University of California.

Voigt, J. (1985). Patterns and routines in classroom interaction. *Recherches en Didactique des Mathématiques, 6*(1), 69–118.

Von Glasersfeld, E. (1995). *Radical Constructivism: A Way of Knowing and Learning*. London: Taylor and Francis.

Walcott, C., Mohr, D. and Kastberg, S. E. (2009). Making sense of shape: An analysis of children's written responses. *Journal of Mathematical Behavior, 28*(1), 30–40.

Wang, S. (2013). Prospective Teachers' learning in geometry: Changes in discourse and thinking. *New Waves-Educational Research & Development, 16*(1), 92–123.

Wang, S. and Kinzel, M. (2014). How do they know it is a parallelogram? Analysing geometric discourse at van Hiele Level 3. *Research in Mathematics Education, 16*(3), 288–305.

Wang, Y. (1999). *Hua Loo-Keng*. Trans. Peter Shiu. Singapore: Springer Verlag.

Wares, A. (2010). Using dynamic geometry to explore non-traditional theorems. *International Journal of Mathematical Education in Science and Technology, 41*(3), 351–358.

Waschkies, H.J. (1989). *Anfänge der Arithmetik im alten Orient und bei den Griechen.* Amsterdam: B. R. Grüner.

Weiss, M. and Herbst, P. (2015). The role of theory building in the teaching of secondary geometry. *Educational Studies in Mathematics, 89*(2), 205–229.

Weiss, M., Herbst, P. and Chen, C. (2009). Teachers' perspectives on "authentic mathematics" and the two-column proof form. *Educational Studies in Mathematics, 70*(3), 275–293.

Whiteley, W. (1999). The decline and rise of geometry in 20th century North America. In J. G. McLoughlin (Ed.), *Canadian Mathematics Education Study Group Conference Proceedings* (pp. 7–30). St John's, NF: Memorial University of Newfoundland.

Widder, M. and Gorsky, P. (2013). How students use a software application for visualizing 3D geometric objects to solve problems. *Journal of Computers in Mathematics and Science Teaching, 32*(1), 89–120.

Wilkins, J. L. and Brand, B. R. (2004). Change in preservice teachers' beliefs: An evaluation of a mathematics methods course. *School Science and Mathematics, 104*(5), 226–232.

Wilkins, J. L. and Norton, A. (2011). The splitting loope. *Journal for Research in Mathematics Education, 42*(4), 386–416.

Willson, W. W. (1977). *The mathematics curriculum: Geometry.* Glasgow: Blackie/Schools Council.

Wittgenstein, L. (1958). *Philosophical investigations* (2nd edition) (translated by G. E. M. Anscombe). Oxford: Blackwell. (Original work published 1953).

Wittmann, E. C. (1995). Mathematics education as a 'design science'. *Educational Studies in Mathematics, 29*(4), 355–374.

Wong, L. (2013). Spaghetti quadrilaterals. *Mathematics Teaching, 232*, 19–20.

Wu, H. (1996). The mathematician and the mathematics education reform. *Notices of the American Mathematical Society 43*(12), 1531–1537.

Wu, H. (2011). The mis-education of mathematics teachers. *Notices of the AMS, 58*(3), 372–384.

Wyller, E. A. (1964). Definition and hypothesis in Plato's Meno. *Inquiry: An Interdisciplinary Journal of Philosophy, 1*(1), 219–226.

Yamamoto, S. (2006). The process of adapting a German pedagogy for modern mathematics teaching in Japan. *Paedagogica Historica, 42*(4–5), 535–545.

Yanik, H. B. (2011). Prospective middle school mathematics teachers' preconceptions of geometric translations. *Educational Studies in Mathematics, 78*(2), 231–260.

Yerushalmy, M. and Houde, R. A. (1986). The Geometric Supposer: Promoting thinking and learning. *The Mathematics Teacher, 79*(6), 418–422.

Zazkis, R. and Leikin, R. (2008). Exemplifying definitions: A case of a square. *Educational Studies in Mathematics, 69*(2), 131–148

Zazkis, R. and Leikin, R. (2010). Advanced mathematical knowledge in teaching practice: Perceptions of secondary mathematics teachers. *Mathematical Thinking and Learning, 12*(4), 263–281.

Zazkis, R. and Mamolo, A. (2011). Reconceptualizing knowledge at the mathematical horizon. *For the Learning of Mathematics, 31*(2), 8–13.

Zur Oeveste, H. (1987). *Kognitive Entwicklung im Vor-und Grundschulalter: eine Revision der Theorie Piagets.* Göttingen: Verlag für Psychologie, Hogrefe.

INDEX

FSC
www.fsc.org
MIX
Papier | Fördert gute Waldnutzung
FSC® C083411

Kolibri 360 GmbH
Ferdinand-Jühlke-Straße 7
99095 Erfurt, Deutschland
produktsicherheit@kolibri360.de